T0255539

Lecture Notes in Computer Science 13648

The series Lecture Notes in Computer Science (LNCS), including its subseries Lecture Notes in Artificial Intelligence (LNAI) and Lecture Notes in Bioinformatics (LNBI), has established itself as a medium for the publication of new developments in computer science and information technology research, teaching, and education.

LNCS enjoys close cooperation with the computer science R & D community, the series counts many renowned academics among its volume editors and paper authors, and collaborates with prestigious societies. Its mission is to serve this international community by providing an invaluable service, mainly focused on the publication of conference and workshop proceedings and postproceedings. LNCS commenced publication in 1973.

Yiming Xiao · Guanyu Yang · Shuang Song
Editors

Lesion Segmentation in Surgical and Diagnostic Applications

MICCAI 2022 Challenges, CuRIOUS 2022, KiPA 2022 and MELA 2022, Held in Conjunction with MICCAI 2022 Singapore, September 18–22, 2022 Proceedings

Editors
Yiming Xiao
Concordia University
Montréal, QC, Canada

Guanyu Yang
Southeast University
Nanjing, China

Shuang Song
Wuhan University
Wuhan, China

ISSN 0302-9743 ISSN 1611-3349 (electronic)
Lecture Notes in Computer Science
ISBN 978-3-031-27323-0 ISBN 978-3-031-27324-7 (eBook)
https://doi.org/10.1007/978-3-031-27324-7

This Springer imprint is published by the registered company Springer Nature Switzerland AG
The registered company address is: Gewerbestrasse 11, 6330 Cham, Switzerland

Kidney Parsing Challenge 2022: Multi-structure Segmentation for Renal Cancer Treatment (KiPA 2022)

Preface

Three-dimensional (3D) kidney parsing on computed tomography angiography (CTA) images is one of the most important tasks for surgery-based renal cancer treatment (e.g., laparoscopic partial nephrectomy). It targets segmenting 3D kidneys, renal tumors, arteries, and veins. Once successful, clinicians will benefit from the 3D visual model of renal structures for accurate preoperative planning. Preoperatively, the renal arteries will help estimate the renal perfusion model, so that the clinicians will select the tumor-feeding arterial branches and locate the arterial clamping position easily. The tumor and kidney models will visually show the lesion regions, thus helping the pre-plan of the tumor resection surface. Intraoperatively, the preoperative plan will be displayed on the screen together with laparoscopic videos to guide the surgery. Renal vessels (veins, arteries) outside the hilum will show a clear arterial clamping region visually, thus the clinicians will select arterial clamping branches quickly. The 3D visual model will also guide the clinicians in making appropriate decisions. Therefore, the costs of treatment will be reduced, the quality of surgery will be improved, and the pain of patients will be relieved. This edition of the KiPA MICCAI challenge provided researchers with an opportunity and common benchmark to characterize their multi-structure segmentation models for kidney parsing, thus exploring solution methods for renal cancer treatment.

After a single-blind peer review with 4 reviews per paper, 6 papers out of 7 submissions have been accepted for publication.

December 2022

Guanyu Yang
Yuting He
Penfei Shao
Xiaomei Zhu

Kidney Parsing Challenge 2022: Multi-Structure Segmentation for Renal Cancer Treatment (KiPA 2022)

Organization

Organization Committee

Guanyu Yang	Jiangsu Provincial Joint International Research Laboratory of Medical Information Processing, Southeast University, China
Yuting He	Jiangsu Provincial Joint International Research Laboratory of Medical Information Processing, Southeast University, China
Penfei Shao	The First Affiliated Hospital of Nanjing Medical University, China
Xiaomei Zhu	The First Affiliated Hospital of Nanjing Medical University, China
Jindi Kong	Jiangsu Provincial Joint International Research Laboratory of Medical Information Processing, Southeast University, China
Lizhan Xu	Jiangsu Provincial Joint International Research Laboratory of Medical Information Processing, Southeast University, China
Ziyue Jiang	Jiangsu Provincial Joint International Research Laboratory of Medical Information Processing, Southeast University, China

Correction of Brain Shift with Intra-Operative Ultrasound Segmentation Challenge (CuRIOUS-SEG 2022)

Preface

Early brain tumor resection can effectively improve the patient's survival rate. However, resection quality and safety can often be heavily affected by intra-operative brain tissue shift due to factors such as gravity, drug administration, intracranial pressure change, and tissue removal. Such tissue shift can displace the surgical target and vital structures (e.g., blood vessels) shown in pre-operative images while these displacements may not be directly visible in the surgeon's field of view. Intra-operative ultrasound (iUS) is a robust and relatively inexpensive technique to track intra-operative tissue shift and surgical tools. Automatic algorithms for brain tissue segmentation in iUS, especially brain tumors and resection cavity, can greatly facilitate the robustness and accuracy of brain shift correction through image registration and allow easy interpretation of the iUS. This has the potential to improve surgical outcomes and patient survival rate. The Correction of Brain Shift with Intra-Operative Ultrasound Segmentation (CuRIOUS-SEG) challenge is an extension to the previous CuRIOUS 2018 & CuRIOUS 2019 Challenge that focused on image registration algorithms. It provides a snapshot of the current and new techniques for iUS segmentation and provides the opportunity to benchmark the methods on the newly released dataset of iUS brain tumor and resection cavity segmentation.

After a single-blind peer review with an average of 2 reviews per paper, 3 papers out of 4 submissions were accepted for publication.

September 2022

Ingerid Reinertsen
Yiming Xiao
Hassan Rivaz
Matthieu Chabanas
Bahareh Behboodi
Francois-Xavier Carton

Organization

Organization Committee

Ingerid Reinertsen	SINTEF, Norway
Yiming Xiao	Concordia University, Canada
Hassan Rivaz	Concordia University, Canada
Matthieu Chabanas	University of Grenoble Alpes, Grenoble Institute of Technology, France
Bahareh Behboodi	Concordia University, Canada
Francois-Xavier Carton	University of Grenoble Alpes, Grenoble Institute of Technology, France

Mediastinal Lesion Analysis Challenge (MELA 2022)

Preface

The mediastinum is the common site of various lesions, including hyperplasia, cysts, tumors, and lymph nodes transferred from the lungs, which might cause serious problems due to their locations. The detection of mediastinal lesions has important indications for the early screening and diagnosis of related diseases. However, it is difficult to distinguish mediastinal lesions in CT scans for clinicians because of image artifact, intensity inhomogeneity, and their similarity with other tissues. Thus, it is necessary to design, implement and validate computer-aided approaches to assist doctors in distinguishing mediastinal lesions from CT scans.

This challenge workshop aimed at bringing together the community of researchers interested in solving the problems of mediastinal lesion detection and medical image analysis based on deep learning. The workshop was co-organized by Wuhan University, China, Shanghai Pulmonary Hospital, China, and Dianei Technology Inc., Shanghai, China, with the support of the JD Explore Academy and the Specialist Alliance of Clinical Skills Promotion and Enhancement for General Thoracic Surgery. The workshop format included a keynote speaker, technical presentations, and a panel. The workshop was attended by around 20 people.

We received five articles for review this year. Each paper secured three reviews from members of the PC. After a thorough peer-reviewing process, we selected one article (an acceptance ratio of 20%) for presentation at the workshop. The review process focused on the quality of the papers, their innovative ideas, and their applicability to the field of medical image analysis.

Finally, we would like to thank all who helped to make the workshop on MELA a success: the authors who submitted papers, the colleagues who refereed the submitted papers, the numerous colleagues who attended the sessions, and the MICCAI 2022 organizers whose invaluable support greatly helped the organization of this workshop.

Organization

Program Committee Chairs

Bo Du — Wuhan University, China
Yong Luo — Wuhan University, China
Jiancheng Yang — Shanghai Jiao Tong University, China

Program Committee

Chang Chen — Shanghai Pulmonary Hospital, China
Xuefei Hu — Shanghai Pulmonary Hospital, China
Kaiming Kuang — Dianei Technology Inc. Shanghai, China
Bingbing Ni — Shanghai Jiao Tong University, China
Shuang Song — Wuhan University, China
Yunlang She — Shanghai Pulmonary Hospital, China
Dong Xie — Shanghai Pulmonary Hospital, China
Rui Xu — Wuhan University, China
Deping Zhao — Shanghai Pulmonary Hospital, China
Mengmeng Zhao — Shanghai Pulmonary Hospital, China
Yuming Zhu — Shanghai Pulmonary Hospital, China

Contents

The 2022 Mediastinal Lesion Analysis Challenge (MELA 2022)

Kidney Parsing Challenge 2022: Multi-Structure Segmentation for Renal Cancer Treatment (KiPA 2022)

A Segmentation Network Based on 3D U-Net for Automatic Renal Cancer Structure Segmentation in CTA Images

Xin Weng, Zuquan Hu, and Fan Yang(✉)

School of Biology and Engineering (School of Health Medicine Modern Industry), Guizhou Medical University, Guiyang, China
yangfan0404@126.com

Abstract. The accuracy segmentation of renal cancer structure (included kidney, renal tumor, renal vein and renal artery) in computed tomography angiography (CTA) images has great clinical significance in clinical diagnosis. In this work, we designed a network architecture based on 3D U-Net and introduced the residual block into network architecture for renal cancer structure segmentation in CTA images. In the network architecture we designed, the multi-scale anisotropic convolution block, dual activation attention block and multi-scale deep supervision equipped for the better segmentation performance. We trained and validated our network in the training set, and tested our network in the opening testing set and closed testing set of KiPA22 challenge. Our method ranked the first place in the KiPA22 challenge leaderboard, and the Hausdorff Distance (HD) of kidney, renal tumor, vein and artery achieved the state-of-the-art, also Dice Similarity Coefficient (DSC) and Average Hausdorff Distance (AVD) of renal artery. According to the results in the KiPA22 challenge, our method have a better segmentation performance in CTA images.

Keywords: Renal cancer segmentation · 3D U-Net · CTA images

1 Introduction

Renal cancer statistics show a small upward trend for both new cases and fatalities associated with the disease since 2012 [1]. Medical imaging have a great importance in the diagnosis and treatment of renal cancer [2]. The segmentation of renal cancer structure in computed tomography angiography (CTA) images can help doctors to observe the kidney and tumor structure, and make a preoperative planning. Therefore, the purpose of Kidney Parsing (KiPA) for Renal Cancer Treatment 2022 Challenge is to develop reliable segmentation methods to promote renal cancer treatment.

With the development of Convolution Neural Networks (CNN), many deep learning based methods were proposed for medical images segmentation tasks. As for volumetric medical images segmentation, 3D U-Net [3] and V-Net [4] were widely used in segmentation tasks. Considering the success of 3D U-Net performed in volumetric medical images segmentation, we proposed a model based on the encoder-decoder architecture of 3D U-Net in this challenge.

Y. Xiao et al. (Eds.): CuRIOUS/KiPA/MELA 2022, LNCS 13648, pp. 3–8, 2023.
https://doi.org/10.1007/978-3-031-27324-7_1

2 Method

2.1 Dataset and Pre-processing

The dataset of KiPA22 challenge[1] was collected from 130 patients' abdominal CTA images [5–8], and the dataset included training set, opening testing set, and closed testing set. The number of training set, opening testing set, and closed testing set is 70, 30, and 30 respectively. Five-fold cross-validation was used in our training phase, and the training set of dataset was divided into training data and validation data randomly, the number of training data and validation data is 56 and 14 respectively. The label of each case has five categories, included background, renal vein, kidney, renal artery, and renal tumor. A case of KiPA22 challenge dataset depicted in Fig. 1.

Several pre-processing operations were conducted by challenge organizers[2], included resampling and cropping operations. Each case of KiPA22 challenge dataset was resampled so that image will have the same voxel spacing in X/Y/Z-axis. Images were cropped the renal region of interesting (ROI) of original CTA images. We also conducted some pre-processing operations for dataset. The intensity values of each case of dataset were adjusted to the range [0, 2048], and normalized by subtracting the mean and dividing by the standard deviation (Z-score normalization).

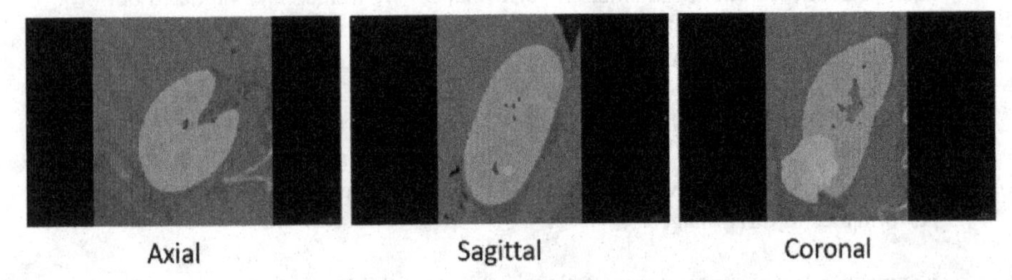

Fig. 1. A representative case in KiPA22 challenge dataset. The region of green, blue, red and yellow stands for kidney, renal artery, renal vein, and renal tumor, respectively.

2.2 Proposed Network Architecture

We proposed a network architecture base 3D U-Net with multi-scale anisotropic convolution block, dual activation attention block and multi-scale deep supervision. The network architecture is depicted in Fig. 2. The network architecture consisted of three major parts, encoder part, multi-scale anisotropic convolution block in the bottom and decoder part. The residual block was introduced in the network, and we modified it as showed in Fig. 2 (where Conv means convolution block, BN means Batch Normalization, and PReLU means Parametric Rectified Linear Unit). In the encoder part, two different residual blocks were used in each layer. The $3 \times 3 \times 3$ convolution with stride of 2 was used for downsampling. The initial number of channels in the first layer was set

[1] https://kipa22.grand-challenge.org/.

[2] https://kipa22.grand-challenge.org/dataset/.

to 30, and be double in each next layer, the maximum number of channels was set to 320. The skip connection was used for concatenating the feature maps from encoder part and decoder part. In the decoder part, transposed convolution was used for upsampling. A residual block was used in each layer of decoder part. Behind the each layer of decoder part, a 1 × 1 × 1 convolution and a softmax layer were used to output the probability maps for multi-scale deep supervision.

We designed the multi-scale anisotropic convolution block and dual activation attention block in the network. Multi-scale anisotropic convolution block with different convolutions was used in the bottom of network to obtain different receptive field of feature maps. Dual activation attention block was used in the end of each layer to catch more important information of feature maps. As for the multi-scale deep supervision, we used the maxpooling operation to obtain the low-resolution ground truths for multi-scale deep supervision.

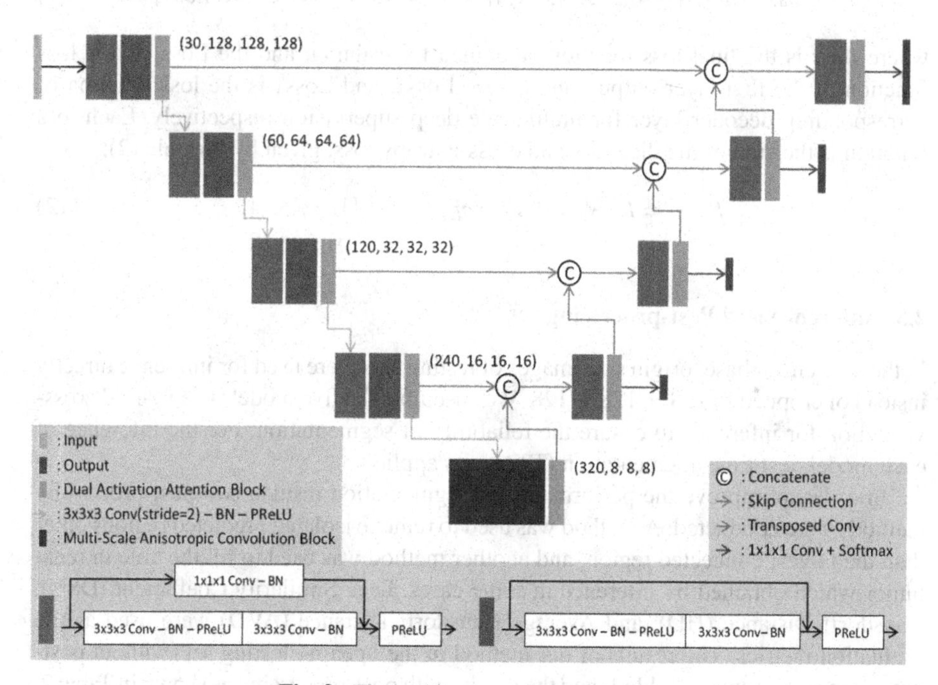

Fig. 2. The architecture of our network.

2.3 Implement Details

Each model of five-fold cross-validation was trained for 150 epochs, and 200 iterations in each epoch. The batch size was set to 2, and patch size was cropped to 128 × 128 × 128. The initial learning rate was set to 0.0003. The cosine decay was applied as the learning rate descent strategy. The model trained with AdamW optimizer with L2 weight decay of 1×10^{-5}. The model with highest mean dice score of four categories (renal vein, kidney, renal artery and renal tumor) performed in validation set was used for inference. Our method was implemented using PyTorch[3] and trained on a computer with a NVIDIA RTX 3090 GPU and 128 GB RAM.

Considering the number of training set and iterations, we used data augmentation during the training phase to avoid overfitting and enhance generalization of model. The data augmentation was applied with the batchgenerators[4] included mirror, elastic deformations, rotation, scale, random crop, Gaussian noise, brightness, Gaussian blur and gamma augmentations.

2.4 Loss Function

The final loss function for training defined as formula (1):

$$Loss = Loss_1 + 0.5 \times Loss_2 + 0.5 \times Loss_3 + 0.5. \times Loss_4 \tag{1}$$

where *Loss* is the final loss function combined for training, and the $Loss_1$ is the loss function of the first layer output, and $Loss_2$, $Loss_3$, and $Loss_4$ is the loss function of corresponding decoder layer for multi-scale deep supervision respectively. Each loss function is the sum of the dice loss and cross-entropy loss, given by formula (2):

$$Loss_i = Loss_{dice} + Loss_{CE}, \quad i \in \{1, 2, 3, 4\} \tag{2}$$

2.5 Inference and Post-processing

In the inference phase, origin size images of testing sets were used for inference directly instead of cropped to 128 × 128 × 128. We ensemble the five models of five-fold cross-validation for inference to ensure the reliability of segmentation. For the inference of each model, test time augmentation (TTA) was applied.

In order to improve the performance of segmentation results, two post-processing methods were conducted. A method was used to remove isolated predicted regions other than the largest connected region, and another method was used to fill the hole in renal tumor which obtained by inference in some cases. Dice Similarity Coefficient (DSC), Hausdorff Distance (HD), and Average Hausdorff Distance (AVD) were used as the evaluation metrics. The results of our method in the opening testing set (without post-processing) as shown in Table 1, and the results with post-processing as shown in Table 2. As the Table 1 and Table 2 described, the post-processing methods we used are effective for segmentation performance and can achieve better performances in some metrics. It has no effect on the segmentation result of kidney, and three metrics of renal artery were further improved. After the post-processing, the DSC of tumor increased, but AVD became worse, and HD of vein had a better performance, but DSC, AVD became worse.

[3] https://pytorch.org/.

[4] https://github.com/MIC-DKFZ/batchgenerators.

Table 1. The results of the opening testing set (without post-processing)

Kidney			Tumor		
DSC	HD	AVD	DSC	HD	AVD
0.9563	17.3545	0.4603	0.8871	11.3319	1.4386
Artery			Vein		
DSC	HD	AVD	DSC	HD	AVD
0.8990	16.7295	0.3573	0.8204	13.3993	1.1291

Table 2. The results of the opening testing set (with post-processing).

Kidney			Tumor		
DSC	HD	AVD	DSC	HD	AVD
0.9563	17.3545	0.4603	0.8874	11.3319	1.4597
Artery			Vein		
DSC	HD	AVD	DSC	HD	AVD
0.9009	15.7704	0.3133	0.8189	13.3748	1.1957

3 Results

The results of the closed testing set in the KiPA22 challenge[5] listed in Table 3, the results included three metrics (DSC, HD, AVD) of four categories (kidney, renal tumor, vein and artery) respectively. It is worth noting that, the HD of kidney, renal tumor, vein and artery achieved the best value by our method in the closed testing set. Besides, the DSC and AVD of renal artery also achieved the best metrics performance compared with other teams.

Table 3. The results of the closed testing set.

Kidney			Tumor		
DSC	HD	AVD	DSC	HD	AVD
0.9594	17.1941	0.4354	0.8846	9.7367	1.2538
Artery			Vein		
DSC	HD	AVD	DSC	HD	AVD
0.9061	10.2650	0.1937	0.8310	11.1366	0.6337

[5] https://kipa22.grand-challenge.org/evaluation/open-evaluation/leaderboard/.

4 Discussion and Conclusion

In this work, we designed a segmentation network based on 3D U-Net for the renal cancer structure segmentation (included kidney, renal tumor, vein and artery) in CTA images. The results performed in the KiPA22 challenge demonstrate the superiority of our method in renal cancer structure segmentation, and the effectiveness of encoder-decoder segmentation network architecture in medical images segmentation tasks.

References

1. Sung, W.W., Ko, P.Y., Chen, W.J., Wang, S.C., Chen, S.L.: Trends in the kidney cancer mortality-to-incidence ratios according to health care expenditures of 56 countries. Sci. Rep. **11**, 1479 (2021). https://doi.org/10.1038/s41598-020-79367-y
2. Rossi, S.H., Prezzi, D., Kelly-Morland, C., Goh, V.: Imaging for the diagnosis and response assessment of renal tumours. World J. Urol. **36**, 1927–1942 (2018). https://doi.org/10.1007/s00345-018-2342-3
3. Çiçek, Ö., Abdulkadir, A., Lienkamp, S.S., Brox, T., Ronneberger, O.: 3D U-Net: learning dense volumetric segmentation from sparse annotation. In: Ourselin, S., Joskowicz, L., Sabuncu, M.R., Unal, G., Wells, W. (eds.) MICCAI 2016. LNCS, vol. 9901, pp. 424–432. Springer, Cham (2016). https://doi.org/10.1007/978-3-319-46723-8_49
4. Milletari, F., Navab, N., Ahmadi, S.A.: V-Net: fully convolutional neural networks for volumetric medical image segmentation. In: 2016 Fourth International Conference on 3D Vision (3DV), pp. 565–571. IEEE (2016)
5. He, Y., et al.: Meta grayscale adaptive network for 3D integrated renal structures segmentation. Med. Image Anal. **71**, 102055 (2021)
6. He, Y., et al.: Dense biased networks with deep priori anatomy and hard region adaptation: semi-supervised learning for fine renal artery segmentation. Med. Image Anal. **63**, 101722 (2020)
7. Shao, P., et al.: Laparoscopic partial nephrectomy with segmental renal artery clamping: technique and clinical outcomes. Eur. Urol. **59**, 849–855 (2011)
8. Shao, P., et al.: Precise segmental renal artery clamping under the guidance of dual-source computed tomography angiography during laparoscopic partial nephrectomy. Eur. Urol. **62**, 1001–1008 (2012)

Boundary-Aware Network for Kidney Parsing

Shishuai Hu[1,2], Zehui Liao[2], Yiwen Ye[2], and Yong Xia[1,2(✉)]

[1] Ningbo Institute of Northwestern Polytechnical University, Ningbo 315048, China
[2] School of Computer Science and Engineering, Northwestern Polytechnical University, Xi'an 710072, China
yxia@nwpu.edu.cn

Abstract. Kidney structures segmentation is a crucial yet challenging task in the computer-aided diagnosis of surgery-based renal cancer. Although numerous deep learning models have achieved remarkable success in many medical image segmentation tasks, accurate segmentation of kidney structures on computed tomography angiography (CTA) images remains challenging, due to the variable sizes of kidney tumors and the ambiguous boundaries between kidney structures and their surroundings. In this paper, we propose a boundary-aware network (BA-Net) to segment kidneys, kidney tumors, arteries, and veins on CTA scans. This model contains a shared encoder, a boundary decoder, and a segmentation decoder. The multi-scale deep supervision strategy is adopted on both decoders, which can alleviate the issues caused by variable tumor sizes. The boundary probability maps produced by the boundary decoder at each scale are used as attention to enhance the segmentation feature maps. We evaluated the BA-Net on the Kidney PArsing (KiPA) Challenge dataset and achieved an average Dice score of 89.65% for kidney structures segmentation on CTA scans using 4-fold cross-validation. The results demonstrate the effectiveness of the BA-Net. Code and pretrained models are available at https://github.com/ShishuaiHu/BA-Net.

Keywords: Boundary-aware network · Medical image segmentation · Kidney parsing

1 Introduction

Accurate kidney-related structures segmentation using computed tomography angiography (CTA) images provides crucial information such as the interrelations among vessels and tumors as well as individual positions and shapes in the standardized space, playing an essential role in computer-aided diagnosis applications such as renal disease diagnosis and surgery planning [3,4,15,16]. Since manual segmentation of kidney structures is time-consuming and requires high concentration and expertise, automated segmentation methods are highly demanded to accelerate this process. However, this task remains challenging due

Y. Xiao et al. (Eds.): CuRIOUS/KiPA/MELA 2022, LNCS 13648, pp. 9–17, 2023.
https://doi.org/10.1007/978-3-031-27324-7_2

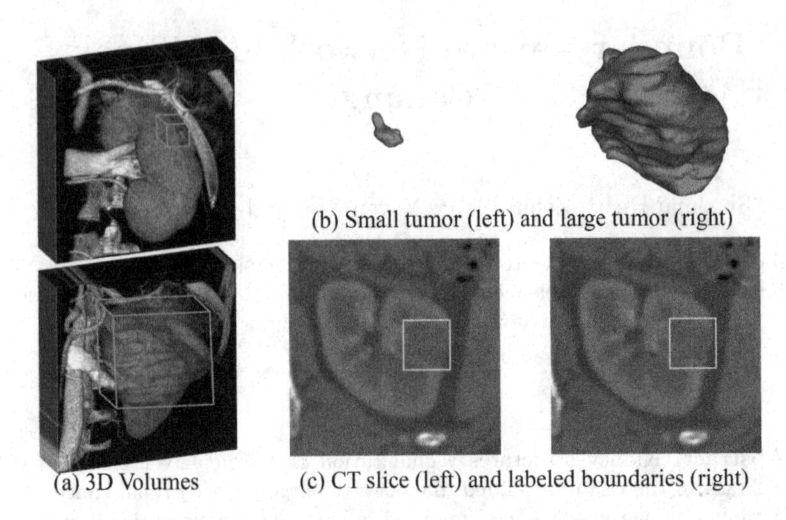

(b) Small tumor (left) and large tumor (right)
(a) 3D Volumes
(c) CT slice (left) and labeled boundaries (right)

Fig. 1. Illustrations of (a) 3D kidney regions of interests (ROIs), (b) imbalanced tumor volumes, and (c) ambiguous boundaries between kidney structures and their surroundings.

to two reasons: (1) the size of different subtypes of kidney tumors, *e.g.*, papillary tumor and clear cell carcinoma, vary significantly in the volumes, as shown in Fig. 1 (b); and (2) the contrast between kidney structures and their anatomical surroundings is particularly low, resulting in the blurry and ambiguous boundary, as shown in Fig. 1 (c).

With the recent success of deep convolutional neural networks (DCNN) in many vision tasks, a number of DCNN-based methods have been proposed for kidney-related structures segmentation. Milletari *et al.* [12] proposed a simple but efficient V-shape network and a Dice loss for volumetric medical image segmentation. Based on that, Isensee *et al.* [8] integrated the data attributes and network design using empirical knowledge and achieved promising performance on the kidney tumor segmentation task [5]. Following the same philosophy, Peng *et al.* [13] and Huang *et al.* [7] searched networks from data directly using NAS and further enhanced the accuracy of abdominal organ segmentation. Although achieved promising performance, these methods [7,8,12,13] ignore the boundary information, which is essential for kidney structure segmentation. Compared to the largely imbalanced volume proportions of different tumors, the boundary (*a.k.a.* surface in 3D) proportions of different tumors are less sensitive to the variable sizes. Therefore, a lot of boundary-involved segmentation methods have been recently proposed for medical image segmentation. Kervadec *et al.* [11] introduced a boundary loss that uses a distance metric on the space of boundaries for medical image segmentation to enhance the segmentation accuracy near the boundary. Karimi *et al.* [10] proposed a Hausdorff distance loss to reduce the hausdorff distance directly to improve the similarity between the predicted mask and ground truth. Shit *et al.* [17] incorporated morphological operation into loss function calculation of thin objects, and proposed a clDice loss for accurate boundary segmentation. Jia *et al.* [9] introduced a 2D auxiliary boundary detection branch that shares the encoder with the 3D segmentation branch and achieved superior performance on MRI prostate segmentation. Despite the improved performance, the boundary-involved loss function-based methods [10,11,17] are time-consuming during training since these boundary-involved loss functions are computing-unfriendly, *i.e.*, the distance map of each training image mask should be computed during each iteration. Whereas the previous auxiliary boundary task-based method [9] does not fully utilize the extracted boundary information since the auxiliary boundary branch will be abandoned at inference time. In our previous work [6], we adopted a 3D auxiliary boundary decoder and introduced the skip connections from the boundary decoder to the segmentation decoder to boost the kidney tumor segmentation performance.

In this paper, we propose a boundary-aware network (BA-Net) based on our previous work [6] for kidney, renal tumor, renal artery, and renal vein segmentation. The BA-Net is an encoder-decoder structure [14]. To force the model to pay more attention to the error-prone boundaries, we introduce an auxiliary boundary decoder to detect kidney-related structures' ambiguous boundaries. Compared to our previous work in [6], both the boundary and segmentation decoders are supervised at each scale of the decoder in this paper to further improve the model's robustness to varying sizes of target organs. Also, we modified the feature fusion mechanism and used the detected boundary probability maps as attention maps on the corresponding segmentation features instead of directly concatenating the boundary feature maps with segmentation feature maps. We have evaluated the proposed BA-Net model on the KiPA Challenge training dataset using 4-fold cross-validation and achieved a Dice score of 96.59% for kidney segmentation, 90.74% for renal tumor segmentation, 87.75% for renal artery segmentation, and 83.53% for renal vein segmentation.

2 Dataset

The kidney-related structures segmentation dataset published by KiPA challenge[1] was used for this study. The KiPA dataset contains 130 CT scans with voxel-level annotations of 4 kidney structures, including the kidney, renal tumor, renal artery and renal vein. Among all the scans, 70 scans are provided as training cases, 30 scans are used for open test evaluation, and the left 30 scans are withheld for close test evaluation. Only the voxel-level annotations of training cases are publicly available, while the annotations of open test cases and close test cases cannot be accessed.

3 Method

The proposed BA-Net contains a shared encoder, a boundary decoder, and a segmentation decoder, as shown in Fig. 2. The encoder extracts features from the

[1] https://kipa22.grand-challenge.org/.

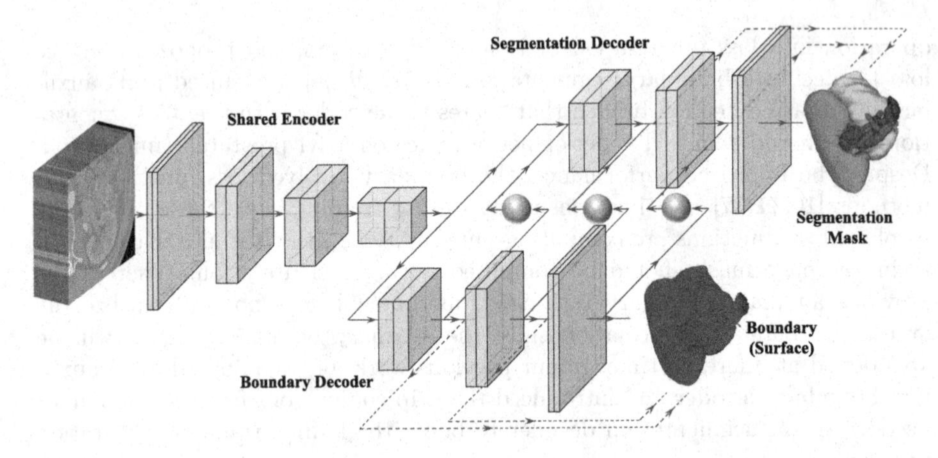

Fig. 2. Illustration of BA-Net for kidney-related structures segmentation. The green, blue, and pink layers represent convolutional layers of the encoder, boundary decoder, and segmentation decoder respectively. The gray nodes represent the soft-max operations. The dashed lines in both decoders represent deep supervision. The skip connections from the shared encoder to both decoders are omitted for simplicity. (Color figure online)

input images. Then the boundary decoder performs boundary detection using the extracted features. The segmentation decoder takes the extracted features as input and outputs segmentation maps of target structures, wherein the boundary probability maps produced by the boundary decoder are adopted as attention maps to enhance the segmentation features.

3.1 Shared Encoder

Since nnUNet [8] performs well on kidney tumor segmentation [5], we adopt it as the backbone of our BA-Net. Therefore, the shared encoder design is the same as nnUNet, which is generated from the statistics of the training data. It contains N encoder blocks ($N = 5$ for kidney structures segmentation task), each of which is composed of 2 convolutional layers. Each convolutional layer is followed by instance normalization and the LeakyReLU activation. The downsample operations are performed using strided convolutions, followed after each encoder block. In the encoder, the number of convolutional filters is set to 32 in the first layer, then doubled in each next block, and finally fixed with 320 when it becomes larger than 256.

3.2 Boundary Decoder

Symmetrically, the boundary decoder contains N decoder blocks to upsample the feature map extracted by the encoder and gradually refine it to generate the boundary map. The transposed convolution with a stride of 2 is used to improve the resolution of input feature maps. The upsampled feature map is concatenated with the low-level feature map extracted from the corresponding encoder block and then fed to the decoder block. The output feature map f_i^b of each decoder block is processed by a convolutional layer and a soft-max layer to generate the boundary probability map p_i^b at each scale i, where $i = 1, 2, \cdots, N-1$.

3.3 Segmentation Decoder

The structure of the segmentation decoder is similar to the boundary decoder, except for an additional step to enhance the segmentation feature map at each scale. In each block of the segmentation decoder, the upsampled feature map $U(f_{i-1}^s)$ is enhanced with the corresponding boundary probability map p_i^b, shown as follows

$$\widehat{U(f_{i-1}^s)} = (1 + p_i^b) * U(f_{i-1}^s) \tag{1}$$

where $\widehat{U(f_{i-1}^s)}$ is the enhanced feature map, and $U(\cdot)$ represents upsample operation.

The enhanced feature map is concatenated with the feature map extracted from the corresponding encoder block before being further processed by the segmentation decoder block. Similar to the boundary decoder, the output feature map f_i^s of each segmentation decoder block is also processed by a convolutional layer and a soft-max layer to generate the segmentation probability map p_i^s at each scale i.

3.4 Training and Inference

The combination of Dice loss and cross-entropy loss is adopted as the objective for both the boundary detection and segmentation tasks. The joint loss at each scale can be calculated as

$$\begin{aligned}\mathcal{L}_i =& 1 - \frac{2\sum_{v=1}^{V} p_{iv} y_{iv}}{\sum_{v=1}^{V} (p_{iv} + y_{iv} + \epsilon)} \\ &- \sum_{v=1}^{V} (y_{iv} \log p_{iv} + (1 - y_{iv}) \log (1 - p_{iv}))\end{aligned} \tag{2}$$

where p_{iv} and y_{iv} denote the prediction and ground truth of the v-th voxel in the output of the i-th decoder block, V represents the number of voxels, and ϵ is a smooth factor to avoid dividing by 0. For the segmentation task, p_i is the segmentation probability map p_i^s, and y_i is the downsampled segmentation ground truth y_i^s. For the boundary detection task, p_i is the boundary probability map p_i^b, and y_i is the boundary y_i^b of the segmentation ground truth y_i^s.

We use deep supervision for both decoders to ensure the boundary can be detected at each scale to enhance the segmentation feature map and improve the model's robustness to targets' sizes. Totally, the loss is defined as

$$\mathcal{L} = \sum_{i=1}^{N-1} \omega_i (\mathcal{L}_i^s + \mathcal{L}_i^b) \tag{3}$$

where ω_i is a weighting vector that enables higher resolution output to contribute more to the total loss [8], $\mathcal{L}_i^s$ is the segmentation loss, and $\mathcal{L}_i^b$ is the boundary detection loss.

During inference, given a test image, the shared encoder extracts features at each scale, and then each boundary decoder block processes these features and produces a boundary probability map. Based on the extracted features and the boundary probability maps, the segmentation decoder is only required to output the segmentation result of the last decoder block.

4 Experiments and Results

4.1 Implementation Details and Evaluation Metric

The CT scans were resampled to a uniform voxel size of 3 mm × 0.6 mm × 0.6 mm. The Hounsfield Unit (HU) values of CT scans were clipped to the range of [918, 1396] according to the data statistics, and then subtracted the mean and divided by the standard deviation. We applied data augmentation techniques, including random cropping, random rotation, random scaling, random flipping, random Gaussian noise addition, and elastic deformation to generate augmented input volumes with a size of 112 × 128 × 128 voxels. Limited to the GPU memory, the batch size was set to 2. The SGD algorithm was used as the optimizer. The initial learning rate lr_0 was set to 0.01 and decayed according to $lr = lr_0 \times (1 - t/T)^{0.9}$, where t is the current epoch and T is the maximum epoch. T was set to 200 for model selection, and 1000 for online evaluation. The whole framework was implemented in PyTorch and trained using a NVIDIA 2080Ti.

The segmentation performance was measured by the Dice Similarity Coefficient (DSC) and Hausdorff Distance (HD). All experiments were performed using 4-fold cross-validation with connected components-based postprocessing.

4.2 Results

We compared our BA-Net with (1) two SOTA medical image segmentation methods: nnUNet and ResUNet [8]; (2) three boundary-involved loss function-based methods: clDice [17], BD Loss [11], and HD Loss [10]; and (3) our previous boundary-involved segmentation approach: BA-Net* [6], as shown in Table 1. It shows that the overall performance of our BA-Net is not only better than nnUNet and ResUNet, but also superior to boundary-involved loss function-based methods. Also, the BA-Net outperforms our previous BA-Net*. It demonstrates that the boundary probability attention map can enhance the segmentation features effectively.

Since more training iterations under strong data augmentation can improve the segmentation performance, we sufficiently retrained nnUNet, ResUNet, and our BA-Net for 1000 epochs. Table 2 shows the performance of these methods. It shows that the performance of the model can be further improved when trained

Table 1. Performance (DSC %, HD voxel) of BA-Net and six competing methods in kidney structures segmentation on KiPA **training** dataset. The models were trained for **200** Epochs. The best results are highlighted with **bold**.

Methods	Kidney		Renal tumor		Renal artery		Renal vein		**Average**	
	DSC ↑	HD ↓	DSC ↑	HD ↓	DSC ↑	HD ↓	DSC ↑	HD ↓	**DSC ↑**	**HD ↓**
nnUNet [8]	96.30	2.44	88.50	6.35	87.35	1.60	82.53	5.63	88.67	4.01
clDice [17]	95.83	3.65	87.64	9.06	86.14	3.50	81.59	7.75	87.80	5.99
BD Loss [11]	96.19	1.95	88.34	7.55	86.90	1.95	82.01	5.77	88.36	4.31
HD Loss [10]	96.12	2.58	88.48	7.22	86.40	2.37	82.10	5.76	88.28	4.48
ResUNet [8]	96.33	2.70	**89.92**	6.70	87.15	2.35	82.63	6.24	89.01	4.50
BA-Net* [6]	96.39	1.87	89.29	5.79	87.23	1.90	**83.55**	6.07	89.12	3.91
BA-Net (Ours)	**96.47**	**1.59**	89.74	**4.78**	**87.59**	**1.25**	83.17	**4.10**	**89.24**	**2.93**

Table 2. Performance (DSC %, HD voxel) of BA-Net and three competing methods in kidney structures segmentation on KiPA **training** dataset. The models were trained for **1000** Epochs. The best results are highlighted with **bold**. 'RB-Ensemble' represent the performance of the averaged result of ResUNet and BA-Net.

Methods	Kidney		Renal tumor		Renal artery		Renal vein		**Average**	
	DSC ↑	HD ↓	DSC ↑	HD ↓	DSC ↑	HD ↓	DSC ↑	HD ↓	**DSC ↑**	**HD ↓**
nnUNet [8]	96.40	1.89	90.23	6.97	87.30	1.63	82.91	5.14	89.21	3.91
ResUNet [8]	96.48	1.79	90.62	5.41	87.33	1.51	82.81	5.37	89.31	3.52
BA-Net	96.54	1.81	90.22	**4.61**	87.36	**1.18**	83.02	4.98	89.28	**3.14**
RB-Ensemble	**96.59**	**1.79**	**90.74**	5.36	**87.48**	1.44	**83.21**	**4.01**	**89.51**	3.15

Table 3. Performance (DSC %, HD *mm*) of BA-Net and four competing methods in kidney structures segmentation on KiPA **open test** dataset. The best results are highlighted with **bold**. 'RB-Ensemble' represent the performance of the averaged result of ResUNet and BA-Net.

Methods	Kidney		Renal tumor		Renal artery		Renal vein		**Average**	
	DSC ↑	HD ↓	DSC ↑	HD ↓	DSC ↑	HD ↓	DSC ↑	HD ↓	**DSC ↑**	**HD ↓**
DenseBiasNet [3]	94.60	23.89	79.30	27.97	84.50	26.67	76.10	34.60	83.63	28.28
MNet [2]	90.60	44.03	65.10	61.05	78.20	47.79	73.50	42.60	76.85	48.87
3D U-Net [1]	91.70	18.44	66.60	24.02	71.90	22.17	60.90	22.26	72.78	21.72
BA-Net	95.87	**16.30**	89.23	12.07	87.53	**14.87**	**85.32**	12.53	89.49	13.94
RB-Ensemble	**95.91**	16.61	**89.64**	**11.52**	**87.78**	15.18	85.30	**12.27**	**89.66**	**13.89**

in more iterations. Still, our BA-Net achieves better performance than nnUNet and is comparable with ResUNet. Therefore, we averaged the results produced by ResUNet and our BA-Net, denoted as RB-Ensemble. It reveals from Table 2 that the averaged results achieved the best DSC, but slightly worse HD than our BA-Net.

We also evaluated the performance of our BA-Net and the ensemble result on the open test set of the KiPA challenge dataset. Table 3 shows the performance of these two methods and three approaches designed for kidney parsing.

It demonstrates that our BA-Net achieves much better performance than the previous kidney structure segmentation methods, confirming the effectiveness of our BA-Net.

5 Conclusion

In this paper, we propose a BA-Net for kidney-related structure segmentation. It is composed of a shared encoder for feature extraction, a boundary decoder for boundary detection, and a segmentation decoder for target organ segmentation. At each scale of the boundary decoder and the segmentation decoder, the boundary and segmentation mask at the corresponding scale are used to supervise the predicted probability map. The predicted boundary map at each decoder block is used as an attention map to enhance the segmentation features. Experimental results on the KiPA challenge dataset demonstrate the effectiveness of our BA-Net.

Acknowledgement. This work was supported in part by the Natural Science Foundation of Ningbo City, China, under Grant 2021J052, in part by the Ningbo Clinical Research Center for Medical Imaging under Grant 2021L003, in part by the Open Project of Ningbo Clinical Research Center for Medical Imaging under Grant 2022LYK-FZD06, and in part by the National Natural Science Foundation of China under Grants 62171377.

References

1. Çiçek, Ö., Abdulkadir, A., Lienkamp, S.S., Brox, T., Ronneberger, O.: 3D U-net: learning dense volumetric segmentation from sparse annotation. In: Ourselin, S., Joskowicz, L., Sabuncu, M.R., Unal, G., Wells, W. (eds.) MICCAI 2016. LNCS, vol. 9901, pp. 424–432. Springer, Cham (2016). https://doi.org/10.1007/978-3-319-46723-8_49
2. Dong, Z., et al.: MNET: rethinking 2d/3d networks for anisotropic medical image segmentation. arXiv preprint arXiv:2205.04846 (2022)
3. He, Y., et al.: Dense biased networks with deep priori anatomy and hard region adaptation: semi-supervised learning for fine renal artery segmentation. Med. Image Anal. **63**, 101722 (2020)
4. He, Y., et al.: Meta grayscale adaptive network for 3d integrated renal structures segmentation. Med. Image Anal. **71**, 102055 (2021)
5. Heller, N., et al.: The state of the art in kidney and kidney tumor segmentation in contrast-enhanced CT imaging: results of the kits19 challenge. Med. Image Anal. **67**, 101821 (2021)
6. Hu, S., Zhang, J., Xia, Y.: Boundary-aware network for kidney tumor segmentation. In: Liu, M., Yan, P., Lian, C., Cao, X. (eds.) MLMI 2020. LNCS, vol. 12436, pp. 189–198. Springer, Cham (2020). https://doi.org/10.1007/978-3-030-59861-7_20
7. Huang, Z., Wang, Z., zhikai yang, Gu, L.: ADWU-Net: adaptive depth and width u-net for medical image segmentation by differentiable neural architecture search. In: Medical Imaging with Deep Learning (2022). https://openreview.net/forum?id=kF-d1SKWJpS

8. Isensee, F., Jaeger, P.F., Kohl, S.A.A., Petersen, J., Maier-Hein, K.H.: nnU-net: a self-configuring method for deep learning-based biomedical image segmentation. Nat. Methods. **18**(2), 203–211 (2021). https://doi.org/10.1038/s41592-020-01008-z https://www.nature.com/articles/s41592-020-01008-z
9. Jia, H., Song, Y., Huang, H., Cai, W., Xia, Y.: HD-Net: hybrid discriminative network for prostate segmentation in MR images. In: Shen, D., et al. (eds.) MICCAI 2019. LNCS, vol. 11765, pp. 110–118. Springer, Cham (2019). https://doi.org/10.1007/978-3-030-32245-8_13
10. Karimi, D., Salcudean, S.E.: Reducing the hausdorff distance in medical image segmentation with convolutional neural networks. IEEE Trans. Med. Imaging **39**(2), 499–513 (2019)
11. Kervadec, H., Bouchtiba, J., Desrosiers, C., Granger, E., Dolz, J., Ayed, I.B.: Boundary loss for highly unbalanced segmentation. In: International Conference on Medical Imaging with Deep Learning, pp. 285–296. PMLR (2019)
12. Milletari, F., Navab, N., Ahmadi, S.A.: V-net: fully convolutional neural networks for volumetric medical image segmentation. In: 2016 Fourth International Conference on 3D Vision (3DV), pp. 565–571. IEEE (2016)
13. Peng, C., et al.: HyperSegNAS: bridging one-shot neural architecture search with 3d medical image segmentation using hypernet. In: Proceedings of the IEEE/CVF Conference on Computer Vision and Pattern Recognition (CVPR), pp. 20741–20751, June 2022
14. Ronneberger, O., Fischer, P., Brox, T.: U-net: convolutional networks for biomedical image segmentation. In: Navab, N., Hornegger, J., Wells, W.M., Frangi, A.F. (eds.) MICCAI 2015. LNCS, vol. 9351, pp. 234–241. Springer, Cham (2015). https://doi.org/10.1007/978-3-319-24574-4_28
15. Shao, P., et al.: Laparoscopic partial nephrectomy with segmental renal artery clamping: technique and clinical outcomes. Eur. Urol. **59**(5), 849–855 (2011)
16. Shao, P., et al.: Precise segmental renal artery clamping under the guidance of dual-source computed tomography angiography during laparoscopic partial nephrectomy. Eur. Urol. **62**(6), 1001–1008 (2012)
17. Shit, S., et al.: CLDICE-a novel topology-preserving loss function for tubular structure segmentation. In: Proceedings of the IEEE/CVF Conference on Computer Vision and Pattern Recognition, pp. 16560–16569 (2021)

A CNN-Based Multi-stage Framework for Renal Multi-structure Segmentation

Yusheng Liu, Zhongchen Zhao, and Lisheng Wang(✉)

Institute of Image Processing and Pattern Recognition, Department of Automation, Shanghai Jiao Tong University, Shanghai 200240, People's Republic of China
{lys_sjtu,13193491346,lswang}@sjtu.edu.cn

Abstract. Three-dimensional (3D) kidney parsing on computed tomography angiography (CTA) images has significant therapeutic implications. 3D visual model of kidney, renal tumor, renal vein and renal artery helps clinicians make accurate preoperative planning. In this paper, we utilize a modified nnU-Net named nnHra-Net network, and propose a multi-stage framework with coarse-to-fine and ensemble learning strategy to precisely segment the multi-structure of kidney. Our method is quantitatively evaluated on a public dataset from MICCAI 2022 Kidney Parsing for Renal Cancer Treatment Challenge (KiPA2022), with mean Dice similarity coefficient (DSC) as 95.91%, 90.65%, 88.60% and 85.36% for the kidneys, kidney tumors, arteries and veins respectively, wining the stable and top performance on both open and close testing in the challenge.

Keywords: nnU-Net · Integrated renal structures segmentation · Kidney cancer

1 Introduction

According to the American Cancer Society, there are about 79,000 new cases of kidney and renal pelvis cancer diagnosed in 2022, and 13,920 deaths [1]. In the treatment of renal tumors, minimally invasive surgical procedures, such as laparoscopic partial nephrectomy (LPN) [2], are currently gaining popularity. Mapping computed tomography angiography (CTA) images to a three-dimensional (3D) kidney model for parsing enables visualization of the lesion region and facilitates effective preoperative planning for selecting tumor-feeding arterial branches and readily locating arterial clamping places [3]. Based on the great clinical significance, MICCAI 2022 Kidney Parsing for Renal Cancer Treatment Challenge (KiPA2022) aims to accelerate the development of reliable, valid, and reproducible methods to reach this need, so as to promote surgery-based renal cancer treatment.

Previous kidney segmentation methods include random forest [4,5], multi-atlases [6], and graph cut [7]. Some [4–6] focus on segmentation of kidney body, and others [7] focus on blood vessels as well as other structures. Recently, many

Y. Xiao et al. (Eds.): CuRIOUS/KiPA/MELA 2022, LNCS 13648, pp. 18–26, 2023.
https://doi.org/10.1007/978-3-031-27324-7_3

deep convolutional neural network (CNN)-based methods [8–10] have shown the potential to segment the kidney and its internal structures (including kidney parenchyma, renal arteries, renal veins and renal mass), and these models can automatically extract features from the data and learn robust and discriminative features. He et al. [11] proposed MGANet to obtain fine-grained feature representation and personalized feature fusion via the selection of grayscale distribution range. Dangi et al. [12] proposed a distance map regularized CNN (DMR-UNet) including two decoder branches to generate the mask and boundary distance maps. However, the lake of correlation between multiple branches results in limited performance improvement.

There are mostly the following difficulties. First, the variable location and appearance of kidney tumor subtypes makes it challenging to acquire fine-grained segmentation from full-sized images. In addition, renal arteries are narrow tubular structures that make up only 0.27 % of the image, resulting in a significant class imbalance issue. Consequently, a precise and effective automatic segmentation method without labeling is urgently needed in clinical practice to alleviate the strain of experts.

In this paper, we utilize a modified nnU-Net named nnHra-Net and propose a multi-stage framework consisting of coarse-to-fine and ensemble learning strategy to efficiently segment kidney structures. The contributions of our work are summarized as follows:

1) We utilize a modified nnU-Net named nnHra-Net, employing two loss functions highly related with hard-region-adaption [13], to adapt the large intra-scale variation, thin structure and small volume ratio of renal arteries.
2) We design a systematic pipeline including several process to make the segmentation more accurate.
3) We propose a multi-stage framework consisting of coarse-to-fine and ensemble learning strategy, which effectiveness and efficiency are demonstrated on KiPA2022 challenge , where we achieve the stable and top performance.

2 Methods

As shown in Fig. 1, the basic framework of our method consists of three main branches aiming at tumor, kidney and vessel respectively. For different structural features, we use different loss functions for supervision. According to the ensemble learning strategy, we then combine the probability maps generated by CNN models to get the fine-grained results. In addition, we design the corresponding systematic pipeline to employ post-processing operations based on image features for each stage.

2.1 Loss Function

As both renal artery and vein account for only a little percent of the whole image resulting in the class imbalance, we decide to use the hard region adaptation (HRA) loss function [13], which will penalize the hard region of the vessels

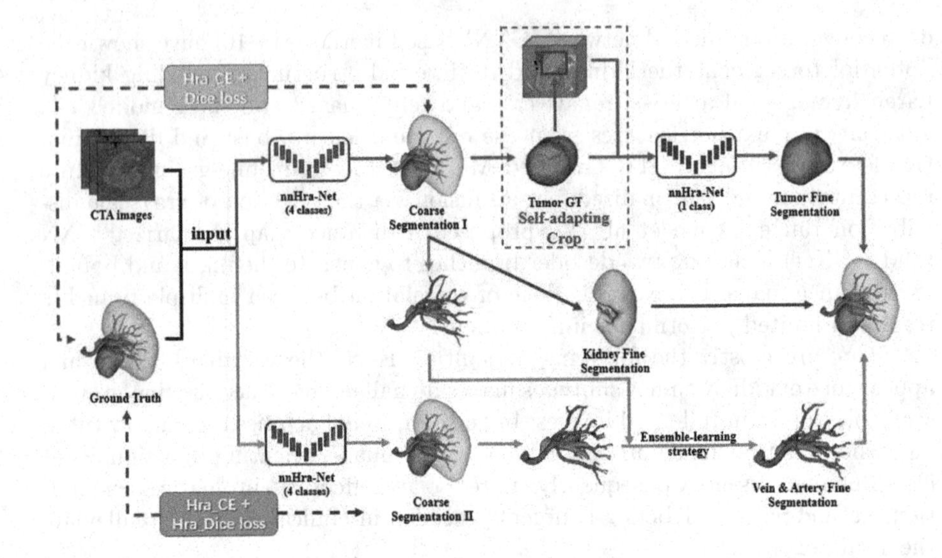

Fig. 1. The overview of our framework. For the branch in blue, we utilize the coarse-to-fine framework and self-adapting crop strategy based on the coarse segmentation to obtain the fine tumor segmentation. For the branch in purple, the kidney fine segmentation are obtained from single stage, namely coarse segmentation I. For the branch in yellow, the artery and vein part are segmented via the ensemble learning strategy from both coarse segmentation I and II (Color figure online).

structure in the image, such as vascular boundary, small ends, etc. [14] and put more emphasis on the boundary of the kidney structure as well.

In this paper, we apply this inter-image sampling strategy to both cross-entropy and dice loss. A detail description of the modified loss functions is as follows.

- The HRA-CE loss function is defined as the special cross-entropy loss, covering the hard-to-segment area instead of all regions, as shown below.

$$L_{HRA-CE} = -\frac{1}{N}\sum_{c=1}^{C}\sum_{n=1}^{N} I(y_{n,c}, \hat{y}_{n,c}) y_{n,c} \log \hat{y}_{n,c}$$

- The HRA-Dice loss function is defined the same as the HRA-CE loss, as shown below.

$$L_{HRA-Dice} = -\frac{1}{N}\sum_{c=1}^{C}\sum_{n=1}^{N} I(y_{n,c}, \hat{y}_{n,c}) \frac{2y_{n,c}\hat{y}_{n,c}}{y_{n,c}^2 \hat{y}_{n,c}^2}$$

where $y_{n,c}$ is the ground truth of the cth class and the nth pixel, while $\hat{y}_{n,c}$ is the predicted value. $I(y_{n,c}, \hat{y}_{n,c})$ is a region-choosing strategy that thresholds for the hard-to-segment mask through the judgement of the L1 distance

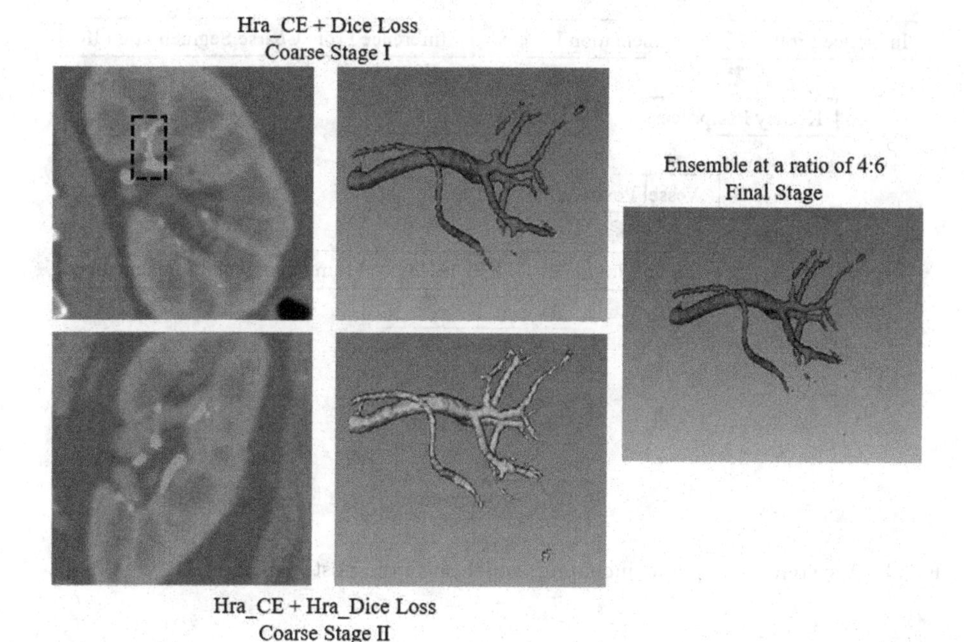

Fig. 2. The light red mask is the result in Coarse Stage I, using Hra-ce and dice loss. The yellow mask is the result in Coarse Stage II with Hra-ce and Hra-dice loss. We employ the ensemble strategy at a ratio of 4:6 to fuse with the final result. (Color figure online)

between the target value and predicted value. If $|y_{n,c} - \hat{y}_{n,c}| < T$, $I(y_{n,c}, \hat{y}_{n,c})$ equals 1, otherwise $I(y_{n,c}, \hat{y}_{n,c})$ equals 0. When T is set as 0, the HRA-CE loss is the same as CE loss.

2.2 Multi-stage Framework

The framework is composed of three stages with a carefully designed nnHra-Net to concentrate on kidney and its internal structures. Specifically, our algorithm contains several emphases:

Fine Kidney Segmentation. As the origin image only contains unilateral kidney with a relatively large ratio, we obtain the predicted mask of kidney from the coarse segmentation I by four-classification nnHra-Net as the fine kidney segmentation.

Fine Tumor Segmentation. We first get the coarse tumor predictions from the coarse stage and then self-adaptingly crop the region of single tumors based on the coarse mask. With the tumor ROI, we re-train a nnHra-Net with the tumor labels only to get the fine tumor segmentation.

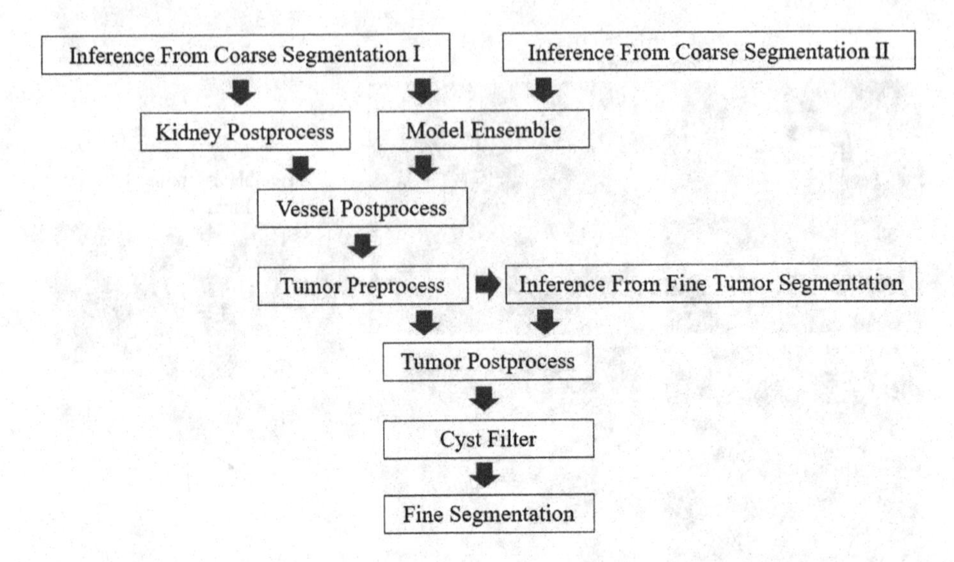

Fig. 3. A systematic pipeline including model selection, postprocessing and cyst filter.

Fine Vein and Artery Segmentation. We use the combination of Hra-ce and dice loss to focus more on large marginal hard-to-learn regions, such as the primary part of the artery and veins, while the combination of Hra-ce and Hra-dice loss maintains the continuity of the thin branches and small ends of vessels from the coronal plane of the image shown in Fig. 2. With several experiments, we use the ensemble strategy at a ratio of 4:6 to fuse with the final result, which can balance the performance on both main and branch part of vessels.

nnHra-Net The proposed nnHra-Net consists of two major parts: a nnU-Net [15] as the baseline, and the hard-region-adaption loss functions. nnU-Net is a standardized self-configuring segmentation benchmark that has won 19 public international competitions, including KiTS2019. Specifically, In the encoder-decoder architecture, each convolutional layer is followed by instance normalization and the LeakyReLU activation.

2.3 Pipeline

As shown in Fig. 3, we design a systematic pipeline including model selection, postprocessing as well as cyst filter. After model inference, we put the predicted results into a systematic process. First, we keep the max connected domain of the kidney. When it comes to vein and artery, we remove the connected domain whose center point distance from the center point of the max connected domain is longer than 100 and 92 respectively, and we also remove the domain with abnormally high CT value for the artery because of the radiographic fact that those are probably the False Postive parts (FPs) where contains other body fluid than blood. As tumors have difference on the CT value [16], we first propose the cyst filter to screen out the potential tumor area via the mean and variance of gray-scale map and then remove the connected domain which are smaller than 100. Finally, we combine the fine kidney, tumor, vein and artery mask together as the fine segmentation.

3 Experiments

3.1 Dataset and Evaluation Metrics

Dataset. We conduct experiments on the 2022 Kidney Parsing for Renal Cancer Treatment Challenge (KiPA2022) dataset [11], which contains abdominal CTA images from patients who have a tumor in their unilateral kidney. For each image, the kidney, renal tumors, arteries, and veins are annotated by mature algorithms, then adjusted by 3 different experts. Pixel sizes of these images are between 0.47 mm/pixel and 0.74 mm/pixel, and their slice's thicknesses are 0.5 mm/pixel. The total number of cases is 130 and 70/30/30 of the original images are selected in Training/Opening test/Closing test phase respectively.

Evaluation Metrics. We calculate metrics for four semantic classes: kidneys, renal tumors, arteries, and veins. Meanwhile, we adopt three evaluation metrics: the Dice similarity coefficient (DSC), the Hausdorff Distance (HD) and the Average Hausdorff Distance (AVD).

3.2 Implementation Details

For image prepropcessing, all the images are resampled to the same target spacing [0.63281, 0.63281, 0.63281] by using the linear spline interpolation so that the resolution of the z-axis is similar to that of the x-/y-axis. A z-score normalization is applied based on the mean and standard deviation of the intensity values from the foreground of the whole dataset. After resampling and normalizing the dataset, we use several data augmentation tricks, including rotation, scaling, mirroring, etc. in the process of training. During the training, the batch size is set as 2 and the batch normalization (BN) is applied. Moreover, we use stochastic gradient descent (SGD) as the optimizer. The initial learning rate is set to be 0.01 and the total training epochs are 300. The patch size is [160, 128, 112] in both the coarse segmentation I & II while [80,80,80] in the fine tumor segmentation stage. We use the 5-fold cross-validation in nnHra-Net to get greater model performance, and other hyper-parameters mainly follow the baseline nnU-Net as default. We implement our network with PyTorch based on a single NVIDIA GeForce RTX 3090 GPU with 24 GB memory.

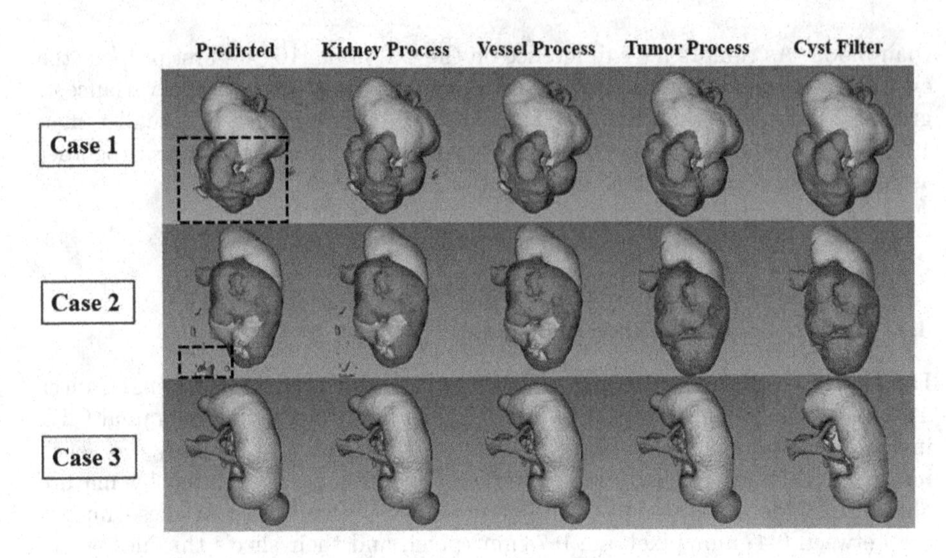

Fig. 4. Visualization of three bad cases through each process in the proposed pipeline.

3.3 Results

To evaluate our method, we randomly divide the training set in KiPA22 challenge which contains 70 cases into training set and test set at a ratio of 60:10. We compare the proposed nnHRA-Net with DenseBiasNet [13], and nnU-Net [15]. Specifically, DenseBiasNet can fuse multi-receptive field and multi-resolution

Table 1. The quantitative evaluation of comparison and ablation study.

Methods	Kidney			Tumor		
	DSC (%)	HD (mm)	AVD (mm)	DSC (%)	HD (mm)	AVD (mm)
DenseBiasNet [13]	94.72	18.51	0.55	74.02	33.97	6.19
nnU-Net [15]	94.96	16.10	0.53	81.94	32.11	3.74
nnHra-Net in kidney Stage	95.09	17.07	0.51	83.33	30.94	3.42
+ Fine tumor stage	95.09	17.07	0.52	84.99	18.23	2.41
+ Fine vessel stage	95.09	17.07	0.51	84.99	30.94	3.42
+ Pipeline (Ours)	**95.16**	**17.01**	**0.48**	**88.25**	**12.66**	**1.53**
Open testing (30 cases)	95.91	15.33	0.45	90.65	8.83	0.89
Close testing (30 cases)	96.02	17.69	0.43	89.61	13.14	1.06
Methods	Vein			Artery		
	DSC(%)	HD(mm)	AVD(mm)	DSC(%)	HD(mm)	AVD(mm)
DenseBiasNet [13]	78.89	17.15	0.93	88.28	16.45	0.32
nnU-Net [15]	79.66	9.71	0.55	84.64	20.22	0.82
nnHra-Net in Kidney Stage	82.77	8.77	0.44	85.60	15.60	0.41
+ Fine Tumor Stage	82.77	8.77	0.44	85.54	15.92	0.41
+ Fine Vessel Stage	83.12	7.23	0.41	86.21	14.97	0.34
+ Pipeline(Ours)	**83.30**	**7.20**	**0.40**	**86.50**	**14.88**	**0.31**
Open Testing(30 cases)	85.36	12.12	0.65	88.60	16.61	0.31
Close Testing(30 cases)	82.88	11.66	0.63	89.12	10.79	0.21

features for a powerful multi-scale representation to accommodate within-scale variations, and nnU-Net is a standardized self-configuring segmentation benchmark that has won 19 public international competitions, including KiTS2019. As shown in Table 1, our network outperforms these two comparative methods on most of the metrics, with DSC of 95.09%, 83.33%, 82.77% and 85.60% in kidney, tumor, vein and artery respectively.

In order to obtain fine-grained segmentation results for each structure, we put nnHra-Net into the proposed muti-stage framework, with corresponding stages and training strategies for kidneys, tumors and vessels. Table 1 shows that our framework performs better on the corresponding structures after adding each stage. With the systematic pipeline, the performance can be further improved with an increase of 4.92% DSC in tumor and 0.9% DSC in renal artery.

There are three cases in the open test which perform bad, and the performances through each process via our pipeline are shown in the Fig. 4. Case 1 and 2 have the same condition that tumor has a extremely large shape which denotes the malignant lesions. As these lesions also have vascular structure, misleading the model to learn those as the artery and vein. What's more, case 2 has the isolated kidney mask in the lower left of the image, which is the spine with a high CT value, resulting in the mistakes of model learning. For Case 3, kidney has both cyst and tumor with similar geometrical characteristic. According to the visualization, these bad cases can be refined through the whole pipeline, demonstrating robustness of our framework.

4 Conclusion

In conclusion, this paper presents a CNN-based multi-stage framework to automatically segment kidneys, kidney tumors, arteries and veins from CTA images. The framework consists of coarse-to-fine stage, ensemble learning strategy and the corresponding loss function focusing hard region. With the proposed systematic pipeline containing several regulations and process on intermediate result, our method achieves the significant improvement with robustness, and wins the stable and top performance on both open and close testing in the challenge.

Acknowledgment. We sincerely appreciate the organizers with the donation of KiPA2022 dataset. The authors of this paper declare that the segmentation method they implemented for participation in the KiPA2022 challenge has not used any pre-trained models nor additional datasets other than those provided by the organizers. The proposed solution is fully automatic without any manual intervention.

References

1. Siegel, R.L., Miller, K.D., Fuchs, H.E., Jemal, A.: Cancer statistics. Cancer J. Clin. **72**(1), 7–33 (2022)
2. Shao, P., et al.: Laparoscopic partial nephrectomy with segmental renal artery clamping: technique and clinical outcomes. Eur. Urol. **59**(5), 849–855 (2011)

3. Shao, P., et al.: Precise segmental renal artery clamping under the guidance of dual-source computed tomography angiography during laparoscopic partial nephrectomy. Eur. Urol. **62**(6), 1001–1008 (2012)
4. Cuingnet, R., Prevost, R., Lesage, D., Cohen, L.D., Mory, B., Ardon, R.: Automatic detection and segmentation of kidneys in 3D CT images using random forests. In: Ayache, N., Delingette, H., Golland, P., Mori, K. (eds.) MICCAI 2012. LNCS, vol. 7512, pp. 66–74. Springer, Heidelberg (2012). https://doi.org/10.1007/978-3-642-33454-2_9
5. Khalifa, F., Soliman, A., Dwyer, A.C., Gimel'farb, G., El-Baz, A.: A random forest-based framework for 3d kidney segmentation from dynamic contrast-enhanced CT images. In: 2016 IEEE International Conference on Image Processing (ICIP), pp. 3399–3403 (2016)
6. Yang, G., et al.: Automatic kidney segmentation in CT images based on multi-atlas image registration. In: 2014 36th Annual International Conference of the IEEE Engineering in Medicine and Biology Society, pp. 5538–5541 (2014)
7. Wang, C., et al.: Precise renal artery segmentation for estimation of renal vascular dominant regions. In: Medical Imaging 2016: Image Processing (M.A. Styner and E.D. Angelini, eds.), vol. 9784, p. 97842M, International Society for Optics and Photonics, SPIE (2016)
8. Çiçek, Ö., Abdulkadir, A., Lienkamp, S.S., Brox, T., Ronneberger, O.: 3D U-net: learning dense volumetric segmentation from sparse annotation. In: Ourselin, S., Joskowicz, L., Sabuncu, M.R., Unal, G., Wells, W. (eds.) MICCAI 2016. LNCS, vol. 9901, pp. 424–432. Springer, Cham (2016). https://doi.org/10.1007/978-3-319-46723-8_49
9. Li, J., Lo, P., Taha, A., Wu, H., Zhao, T.: Segmentation of renal structures for image-guided surgery. In: Frangi, A.F., Schnabel, J.A., Davatzikos, C., Alberola-López, C., Fichtinger, G. (eds.) MICCAI 2018. LNCS, vol. 11073, pp. 454–462. Springer, Cham (2018). https://doi.org/10.1007/978-3-030-00937-3_52
10. Liu, Y., Zhao, Z., Wang, L.: A Multi-Stage Framework for the 2022 Multi-Structure Segmentation for Renal Cancer Treatment. arXiv e-prints. arXiv:2207.09165, July 2022
11. He, Y., et al.: Meta grayscale adaptive network for 3d integrated renal structures segmentation. Med. Image Anal. **71**, 102055 (2021)
12. Dangi, S., Linte, C.A., Yaniv, Z.: A distance map regularized CNN for cardiac cine MR image segmentation. Med. Phys. **46**(12), 5637–5651 (2019)
13. He, Y., et al.: Dense biased networks with deep priori anatomy and hard region adaptation: semi-supervised learning for fine renal artery segmentation. Med. Image Anal. **63**, 101722 (2020)
14. Shit, S., Paetzold, J.C., Sekuboyina, A., Zhylka, A., Menze, B.H.: clDice - a topology-preserving loss function for tubular structure segmentation (2020)
15. Isensee, F., Petersen, J., Kohl, S.A.A., Jäger, P.F., Maier-Hein, K.H.: nnu-Net: breaking the spell on successful medical image segmentation. CoRR, vol. abs/1904.08128 (2019)
16. McClennan, B., Stanley, R., Melson, G., Levitt, R., Sagel, S.: CT of the renal cyst: is cyst aspiration necessary? Am. J. Roentgenol. **133**(4), 671–675 (1979). PMID: 114010

CANet: Channel Extending and Axial Attention Catching Network for Multi-structure Kidney Segmentation

Zhenyu Bu, Kaini Wang, and Guangquan Zhou(✉)

Monash University, Southeast University, School of Biological Science and Medical Engineering, Nanjing, China
guangquan.zhou@seu.edu.cn

Abstract. Renal cancer is one of the most prevalent cancers worldwide. Clinical signs of kidney cancer include hematuria and low back discomfort, which are quite distressing to the patient. Some surgery-based renal cancer treatments like laparoscopic partial nephrectomy relys on the 3D kidney parsing on computed tomography angiography (CTA) images. Many automatic segmentation techniques have been put forward to make multi-structure segmentation of the kidneys more accurate. The 3D visual model of kidney anatomy will help clinicians plan operations accurately before surgery. However, due to the diversity of the internal structure of the kidney and the low grey level of the edge. It is still challenging to separate the different parts of the kidney in a clear and accurate way. In this paper, we propose a channel extending and axial attention catching Network(CANet) for multi-structure kidney segmentation. Our solution is founded based on the thriving nn-UNet architecture. Firstly, by extending the channel size, we propose a larger network, which can provide a broader perspective, facilitating the extraction of complex structural information. Secondly, we include an axial attention catching(AAC) module in the decoder, which can obtain detailed information for refining the edges. We evaluate our CANet on the KiPA2022 dataset, achieving the dice scores of 95.8%, 89.1%, 87.5% and 84.9% for kidney, tumor, artery and vein, respectively, which helps us get fourth place in the challenge.

Keywords: Multi-structure segmentation · Channel extending · Axial attention catching

1 Introduction

Renal cancer is one of the most prevalent cancers in the world, ranking 13th globally and having caused more than 330,000 new cases worldwide. [2] Laparoscopic partial nephrectomy, a dependable and successful treatment for kidney cancer, has gained widespread use. [12] However, laparoscopic partial nephrectomy requires a well-defined segmented anatomy of the kidney. [7,15,16] Since annotating kidney structures manually initially needs trained professionals, which is

Y. Xiao et al. (Eds.): CuRIOUS/KiPA/MELA 2022, LNCS 13648, pp. 27–35, 2023.
https://doi.org/10.1007/978-3-031-27324-7_4

both time-consuming and costly. It is essential to develop a technique for automatic multi-structure kidney segmentation.

Nowadays, numerous deep learning techniques have been developed for medical image segmentation, localization and detection. [1,14,20,21] Since the diversity in the internal structure of the kidney (see in Fig. 1), it it difficult to clearly segment the structure of the kidney to a certain extent. Meanwhile, the low gray level of the edge is also a problem. [7] There are numerous techniques for segmenting the kidney's structure. [5,9,11,18] However, these methods only target a single region, such as the kidney, and do not segment the four components concurrently. This cannot supply sufficient information for the downstream tasks. There are progressive multi-structured divisional approaches, however their effects vary and they are not favorable [10,17].

In this paper, we propose a simple and effective channel extending and axial attention catching Network(CANet) based on the nn-UNet backbone for multi-structure kidney segmentation. Straightforwardly, we enlarge the size of the filters along with adding axial attention module in the decoder phase. Meanwhile, some basic setting of nn-UNet has been modified to get the final results. We use kidney multi-structure data (KiPA22 dataset [6,7,15,16]) to evaluate the proposed method. We demonstrate experimentally that despite the fact that our model invariably increases training time, it generates substantial benefits in our experiment.

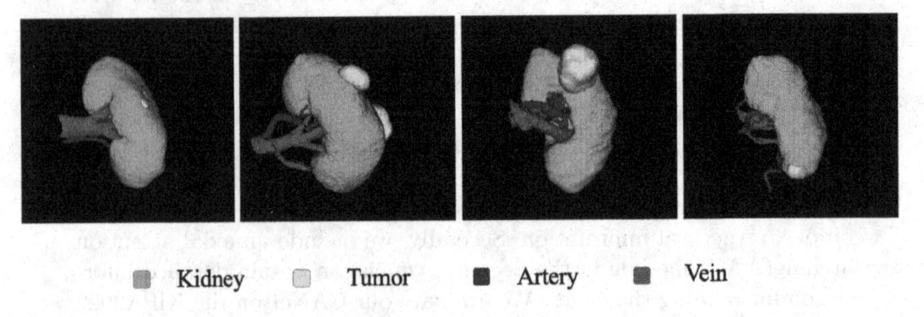

Fig. 1. The shape of the kidneys, tumors, and the fact that the blood vessels aren't very thick are just some of the problems shown in the figure above. The yellow ones are tumors, the green ones are kidneys, and the red and blue ones are veins and arteries. (Color figure online)

2 Methods

Figure 2 illustrates the pipeline of our proposed CANet. First, we preprocess the data, then we transform the input 3D data into the corresponding form. Then we feed the processed data into our CANet, and finally we obtain the final result using post-processing technology.

2.1 Preprocessing

The preprocessing method uses resample and normalization. There are data with different spacing, which is automatically normalized to the median spacing of all

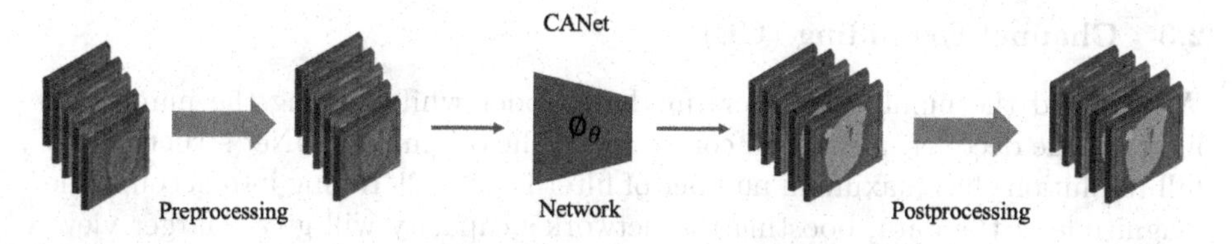

Fig. 2. An overall pipeline of our proposed CANet. This pipeline mainly includes two parts, pre/post processing and the core deep learning network.

data spacing by default. The original data uses third-order spline interpolation and mask uses nearest neighbor interpolation. We count the HU value range of the pixels in the mask in the entire data set and then clip out the HU value range of the [0.05, 99.5] percentage range. Finally the z-score method for normalization is utilized. After these steps, the standard input of nn-UNet is able to be fed to the model for training.

$$z = \frac{x - \mu}{\sigma} \tag{1}$$

where μ is the mean and σ is the variance. It can convert data of different magnitudes into a unified Z score for comparison.

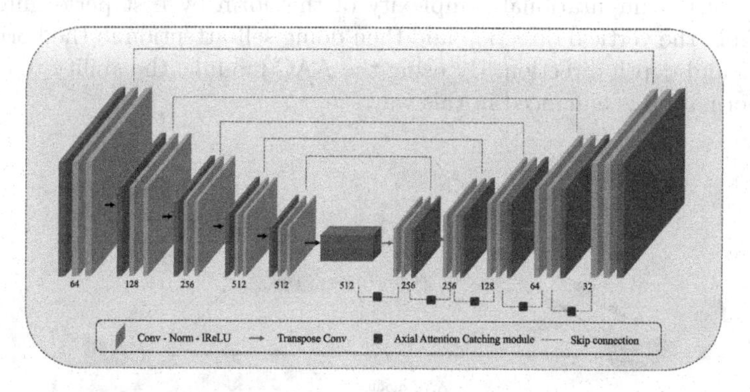

Fig. 3. An overview of our proposed CANet, which is based on the 3D full resolution U-net for kidney multi-structure segmentation.

2.2 nn-UNet Backbone

The 3D full resolution nn-UNet is used as our backbone [8]. The network's structure is encoder-decoder with a skip connection. No new structures, such as residual block, dense connection, or attention mechanism, are added to the original UNet. Instead, it emphasized data preprocessing, training, inference and post-processing.

2.3 Channel Extending (CE)

We doubled the number of filters in the encoder while leaving the number of filters in the decoder unchanged compared to the original nn-UNet structure(3D full resolution) has maximum number of filter is 320. [8] Taking into account the magnitude of the data, boosting the network's capacity will give a larger view, hence simplifying the extraction of intricate structural information. The maximum number of filter is increased to 512. This work was first used in BraTS21 for brain tumor segmentation by Huan Minh Luu. [13] We do not apply group normalization in comparison to their work, instead.

2.4 Axial Attention Catching Module (AAC)

We insert axial attention catching module (AAC module) in the decoder to enhance the representation of the edge feature in order to extract sharper edge information. Figure 4 demonstrates that following the transposed convolution, position embedding is performed on the input, followed by vertical, horizontal, and depth operations, and then the output is concatenated. AAC module is a technique that helps lower the computational complexity [19] along with capturing the global information. For more specific, the axial-attention method minimizes the computational complexity of this form by first performing self-attention in the vertical direction and then doing self-attention in the horizontal direction and depth direction. By using the AAC module, the ability to extract edge information is enhanced in this way.

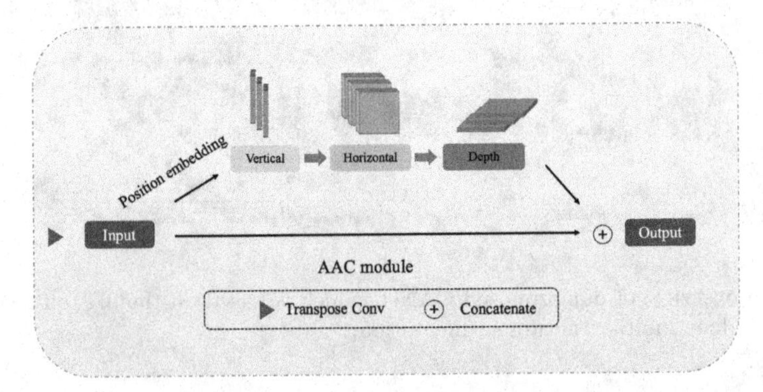

Fig. 4. Visualization of axial attention catching(AAC) module. After transposed convolution, the input has gone through three self-attentions in vertical, horizontal and depth, and finally concatenate to get the final result.

2.5 Loss Function

In our CANet, we use a combination of dice loss and cross-entropy loss as our loss function.

$$\mathcal{L}_{total} = \mathcal{L}_{dice} + \mathcal{L}_{CE} \tag{2}$$

The first item is dice loss. We can calculate the dice coefficient first.

$$DSC\left(y, y'\right) = \frac{2 * |y \cap y'| + \text{ smooth}}{|y| + |y'| + \text{ smooth}} \tag{3}$$

Then we can use the formula below to achieve the final Dice loss.

$$\mathcal{L}_{dice} = 1 - DSC \tag{4}$$

where y indicates the ground truth and $y^{'}$ represents the predicted result.

Cross-entropy loss is represented by the following formula.

$$\mathcal{L}_{CE} = -[y \log \hat{y} + (1 - y) \log(1 - \hat{y})] \tag{5}$$

where y and $y^{'}$ can be considered as ground truth and predict results respectively.

3 Experiments

3.1 Data Preparation

In Kidney Parsing Challenge 2022, original images are CT scans from 130 patients (70 for training, 30 for opening test and 30 for closed test). [6,7,15,16] All of these data was acquired on a Siemens dual-source 64-slice CT scanner and all of these data is in *nii.gz* format. A sample corresponds to a ground truth. The kidneys range in size from 73.73 ml to 263.02 ml, whereas the tumors range from 2.06 ml to 646.68 ml. Meanwhile, there are also 30 samples for the open test phase and 30 samples for the closed test phase.

3.2 Implementation Details

In the training stage, we divide the 70 training data into two parts, 56 for training and the following 14 for validation. Before training, data undergoes preprocessing, such as scaling, random rotation, elastic deformation, gamma scaling, etc. We set the batch size to 2. We employ 300 epochs for each training fold as opposed to 1000 epochs in the origin nn-UNet, which saves a great deal of time and yields excellent results. In our experiment, our training are on NVIDIA RTX 3090 GPU with 24 GB VRAM, which takes about 3 min for one epoch. Finally, we also employ 5-fold cross-validation to improve performance. At the inference stage, we first do the preprocessing as training, and then fed the input into the model directly. After that, we also run the postprecessing method to link some unconnected areas.

4 Results and Discussions

4.1 Ablation Experiments

We do some ablation experiments to confirm the utility of channel extending mechanism and the AAC module in our CANet. In order to perform a fair experiment, all ablation experiments are conducted under same experiment setup.

Here, CE refers to the channel extending mechanism and AAC means axial attention catching module. We test only CE and only AAC separately, and then combine them for testing. As shown in Table 1, the segmentation results were evaluated by dice score. The results show that both of our contributions are effective for improving the performance of the segmentation task.

Table 1. Ablation experiments examine the significance of our discoveries. The channel extension mechanism and AAC module have been shown to be beneficial in the segmentation of several kidney structures.

CE	AAC	DSC(%)			
		Kidney	Tumor	Artery	Vein
✓		95.8	88.6	87.5	84.5
	✓	95.7	88.7	87.3	84.8
✓	✓	95.8	89.1	87.5	84.9

Table 1 presents the quantitative results for the multi-structure kidney segmentation results on four parts, including kidney, tumor, artery and vein. It demonstrates that the channel extending mechanism can get 0.4% and AAC module are effective. Compared to the nn-UNet backbone, the channel extending mechanism can get 0.4%, 0.7%, 0.4% dice score improvement on kidney, artery and vein respectively. Meanwhile, the AAC module can outperform 0.3%, 0.6%, 0.7% dice score on kidney, artery and vein respectively. When combining both of them, the final CANet can get 0.4%, 0.2%, 0.8%, 0.8% dice score on kidney, tumor, artery and vein respectively. These data demonstrate the effectiveness of our contributions.

4.2 Comparison Experiments

We also compare our method with several classical deep learning methods like DenseBiasNet, MNet, 3D Unet. Table 2 demonstrates the quantitative results of our comparison experiments. The number displayed in bold signifies the highest performance in each respective column. In almost every area, our model has the best performance.

Table 2 demonstrates the qualitative results of our proposed CANet compared to three traditional methods and the nn-UNet backbone. From Fig. 2, we can conclude that CANet has the best performance in all results. With regard to DSC, our CANet and nn-UNet backbone have shown excellent performance compared with the other three models. Meanwhile, CANet has improved the results of these four segmentation parts compared with the nn-UNet backbone for 0.4%, 0.2%, 0.8% and 0.8%. The improvement in veins and arteries is still very obvious. Among them, comparing CANet and nn-UNet backbone, our model has a large improvement in $HD(mm)$, which are respectively improved by 1.63 mm, 4.46 mm, 8.12 mm and 8.17 mm.

Table 2. The quantitative assessment reveals that our CANet is superior for our 3D multi-structure kidney segmentation task. On all four renal structures, our CANet outperforms the comparative approaches (DenseBiasNet [5], MNet [4], 3D U-Net [3], and nn-UNet [8]).

Methods	Kidney			Tumor		
	$DSC(\%)$	$HD(mm)$	$AVD(mm)$	$DSC(\%)$	$HD(mm)$	$AVD(mm)$
DenseBiasNet	94.6	23.89	0.79	79.3	27.97	4.33
MNet	90.6	44.03	2.16	65.1	61.05	10.23
3D U-Net	91.7	18.44	0.75	66.6	24.02	4.45
nn-UNet	95.4	18.56	0.78	88.9	14.68	1.50
CANet	**95.8**	**16.93**	**0.46**	**89.1**	**10.22**	**1.23**
Methods	Artery			Vein		
	$DSC(\%)$	$HD(mm)$	$AVD(mm)$	$DSC(\%)$	$HD(mm)$	$AVD(mm)$
DenseBiasNet	84.5	26.67	1.31	76.1	34.60	2.08
MNet	78.2	47.79	2.71	73.5	42.60	3.06
3D U-Net	71.9	22.17	1.10	60.9	22.26	3.37
nn-UNet	86.7	23.24	0.65	84.1	20.04	1.18
CANet	**87.5**	**15.12**	**0.34**	**84.9**	**11.87**	**0.67**

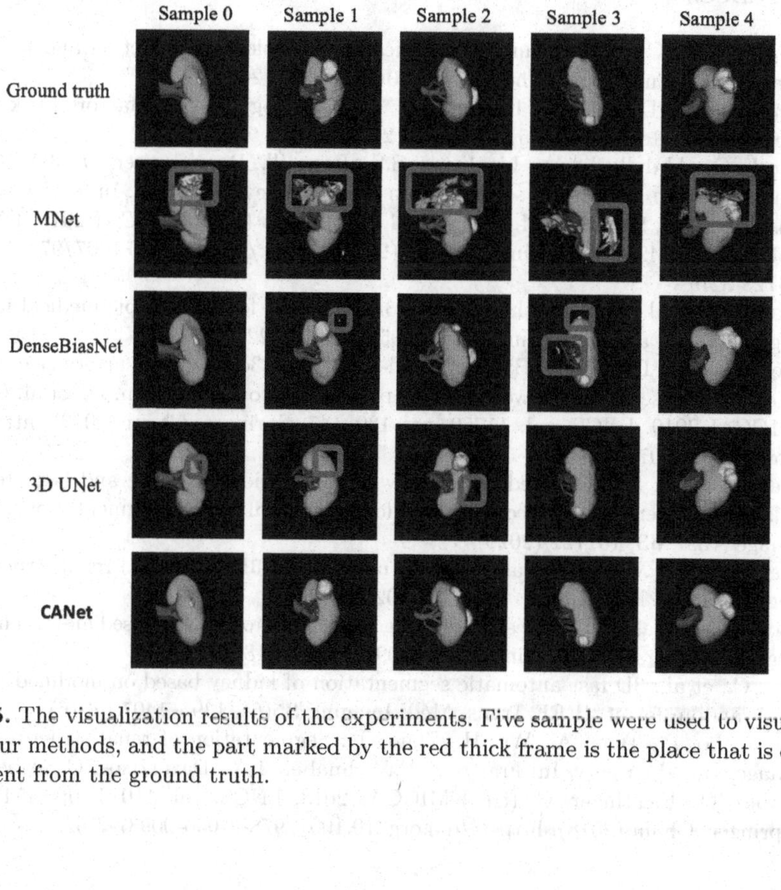

Sample 0 Sample 1 Sample 2 Sample 3 Sample 4

Ground truth

MNet

DenseBiasNet

3D UNet

CANet

Fig. 5. The visualization results of the experiments. Five sample were used to visualize the four methods, and the part marked by the red thick frame is the place that is quite different from the ground truth.

The findings of the visualization are displayed in Fig. 5 below. The other three deep learning approaches do poorly, roughly speaking. Based on the findings of the onion visualization, MNet performs the poorest. DenseBiasNet has the potential to discover more malignancies, but 3D UNet frequently misses tumors. In comparison to existing methods of deep learning, our CANet has obtained superior results in terms of both stereotypes and quantification.

5 Conclusion

This paper proposes a simple but effective method named channel extending and axial attention catching Network(CANet) for multi-structure kidney segmentation based on the nn-UNet framework. To improve the performance of the four-item mixed structure, we begin by employing a channel extending method to assist the extraction of complicated structural information. To further improve the edges, we add an axial attention catching module(AAC) to the decoder. After testing on the KiPA22 dataset, the persuasive findings imply that the newly suggested CANet is successful.

References

1. Cheng, M.M.C., et al.: Nanotechnologies for biomolecular detection and medical diagnostics. Curr. Opin. Chem. Biol. **10**(1), 11–19 (2006)
2. Chow, W.H., Dong, L.M., Devesa, S.S.: Epidemiology and risk factors for kidney cancer. Nat. Rev. Urol. **7**(5), 245–257 (2010)
3. Çiçek, Ö., Abdulkadir, A., Lienkamp, S.S., Brox, T., Ronneberger, O.: 3D U-Net: learning dense volumetric segmentation from sparse annotation. In: Ourselin, S., Joskowicz, L., Sabuncu, M.R., Unal, G., Wells, W. (eds.) MICCAI 2016. LNCS, vol. 9901, pp. 424–432. Springer, Cham (2016). https://doi.org/10.1007/978-3-319-46723-8_49
4. Dong, Z., et al.: MNet: rethinking 2D/3D networks for anisotropic medical image segmentation. arXiv preprint arXiv:2205.04846 (2022)
5. He, Y., et al.: DPA-DenseBiasNet: semi-supervised 3d fine renal artery segmentation with dense biased network and deep priori anatomy. In: Shen, D., et al. (eds.) MICCAI 2019. LNCS, vol. 11769, pp. 139–147. Springer, Cham (2019). https://doi.org/10.1007/978-3-030-32226-7_16
6. He, Y., et al.: Dense biased networks with deep priori anatomy and hard region adaptation: semi-supervised learning for fine renal artery segmentation. Med. Image Anal. **63**, 101722 (2020)
7. He, Y., et al.: Meta grayscale adaptive network for 3D integrated renal structures segmentation. Med. Image Anal. **71**, 102055 (2021)
8. Isensee, F., et al.: nnU-Net: self-adapting framework for U-Net-based medical image segmentation. arXiv preprint arXiv:1809.10486 (2018)
9. Jin, C., et al.: 3D fast automatic segmentation of kidney based on modified AAM and random forest. IEEE Trans. Med. Imaging **35**(6), 1395–1407 (2016)
10. Li, J., Lo, P., Taha, A., Wu, H., Zhao, T.: Segmentation of renal structures for image-guided surgery. In: Frangi, A.F., Schnabel, J.A., Davatzikos, C., Alberola-López, C., Fichtinger, G. (eds.) MICCAI 2018. LNCS, vol. 11073, pp. 454–462. Springer, Cham (2018). https://doi.org/10.1007/978-3-030-00937-3_52

11. Lin, D.T., Lei, C.C., Hung, S.W.: Computer-aided kidney segmentation on abdominal CT images. IEEE Trans. Inf Technol. Biomed. **10**(1), 59–65 (2006)
12. Ljungberg, B., et al.: European association of urology guidelines on renal cell carcinoma: the 2019 update. Eur. Urol. **75**(5), 799–810 (2019)
13. Luu, H.M., Park, S.H.: Extending nn-UNet for brain tumor segmentation. In: Crimi, A., Bakas, S. (eds.) Brainlesion: Glioma, Multiple Sclerosis, Stroke and Traumatic Brain Injuries. BrainLes 2021. Lecture Notes in Computer Science, vol. 12963, pp. pp. 173–186. Springer, Cham (2022). https://doi.org/10.1007/978-3-031-09002-8_16
14. Ronneberger, O., Fischer, P., Brox, T.: U-Net: convolutional networks for biomedical image segmentation. In: Navab, N., Hornegger, J., Wells, W.M., Frangi, A.F. (eds.) MICCAI 2015. LNCS, vol. 9351, pp. 234–241. Springer, Cham (2015). https://doi.org/10.1007/978-3-319-24574-4_28
15. Shao, P., et al.: Laparoscopic partial nephrectomy with segmental renal artery clamping: technique and clinical outcomes. Eur. Urol. **59**(5), 849–855 (2011)
16. Shao, P., et al.: Precise segmental renal artery clamping under the guidance of dual-source computed tomography angiography during laparoscopic partial nephrectomy. Eur. Urol. **62**(6), 1001–1008 (2012)
17. Taha, A., Lo, P., Li, J., Zhao, T.: Kid-Net: convolution networks for kidney vessels segmentation from CT-volumes. In: Frangi, A.F., Schnabel, J.A., Davatzikos, C., Alberola-López, C., Fichtinger, G. (eds.) MICCAI 2018. LNCS, vol. 11073, pp. 463–471. Springer, Cham (2018). https://doi.org/10.1007/978-3-030-00937-3_53
18. Wang, C., et al.: Tensor-cut: a tensor-based graph-cut blood vessel segmentation method and its application to renal artery segmentation. Med. Image Anal. **60**, 101623 (2020)
19. Wang, H., Zhu, Y., Green, B., Adam, H., Yuille, A., Chen, L.-C.: Axial-DeepLab: stand-alone axial-attention for panoptic segmentation. In: Vedaldi, A., Bischof, H., Brox, T., Frahm, J.-M. (eds.) ECCV 2020. LNCS, vol. 12349, pp. 108–126. Springer, Cham (2020). https://doi.org/10.1007/978-3-030-58548-8_7
20. Wang, K.N., et al.: AWSnet: an auto-weighted supervision attention network for myocardial scar and edema segmentation in multi-sequence cardiac magnetic resonance images. Med. Image Anal. **77**, 102362 (2022)
21. Zhou, G.Q., et al.: Learn fine-grained adaptive loss for multiple anatomical landmark detection in medical images. IEEE J. Biomed. Health Inform. **25**(10), 3854–3864 (2021)

Automated 3D Segmentation of Renal Structures for Renal Cancer Treatment

Md Mahfuzur Rahman Siddiquee[1,2], Dong Yang[1], Yufan He[1], Daguang Xu[1], and Andriy Myronenko[1](✉)

[1] School of Computing and AI, Arizona State University, Tempe, AZ, USA
amyronenko@nvidia.com
[2] NVIDIA, Santa Clara, CA, USA

Abstract. The KiPA 2022 challenge aims to develop reliable and reproducible methods for segmenting four kidney-related structures on CTA images to improve surgery-based renal cancer treatment. In this work, we describe our approach for 3D segmentation of renal vein, kidney, renal artery, and tumor from the KiPA 2022 dataset. We used the MONAI framework and the SegResNet architecture with a combination of Dice Focal loss and L2 regularization. We applied various data augmentation techniques and trained the model using the AdamW optimizer with a Cosine annealing scheduler. Our method achieved 10th position in the open test leaderboard and 6th position in the closed test leaderboard.

Keywords: Renal cancer · 3D segmentation · KiPA 2022 challenge

1 Introduction

Renal cancer is a type of cancer that affects the kidneys, a pair of organs in the abdomen that filter waste products from the blood and produce urine. Renal cancer is relatively rare but can be aggressive and spread to other body parts if not treated promptly. Early detection and treatment of renal cancer are important to improve the chances of successful treatment and survival. Some common symptoms of renal cancer include blood in the urine, a lump in the abdomen, and weight loss. Treatment options for renal cancer may include surgery, radiation therapy, and chemotherapy, depending on the stage and type of cancer.

In renal cancer treatment, segmentation identifies and isolates the kidneys and tumors in medical images, such as CT or MRI scans. This information is important for surgical planning and treatment decisions, as it provides detailed information about the location and size of the structures. Segmentation can be performed manually by trained radiologists or by using automated algorithms, such as deep learning models. Automated segmentation can improve the accuracy and reliability of renal cancer treatment by providing detailed, high-resolution information about the structures of interest.

The KiPA 2022 challenge aims to develop reliable and reproducible methods for segmenting four kidney-related structures on CTA images to improve surgery-based renal cancer treatment. The target structures are the kidney, tumor, renal

Y. Xiao et al. (Eds.): CuRIOUS/KiPA/MELA 2022, LNCS 13648, pp. 36–42, 2023.
https://doi.org/10.1007/978-3-031-27324-7_5

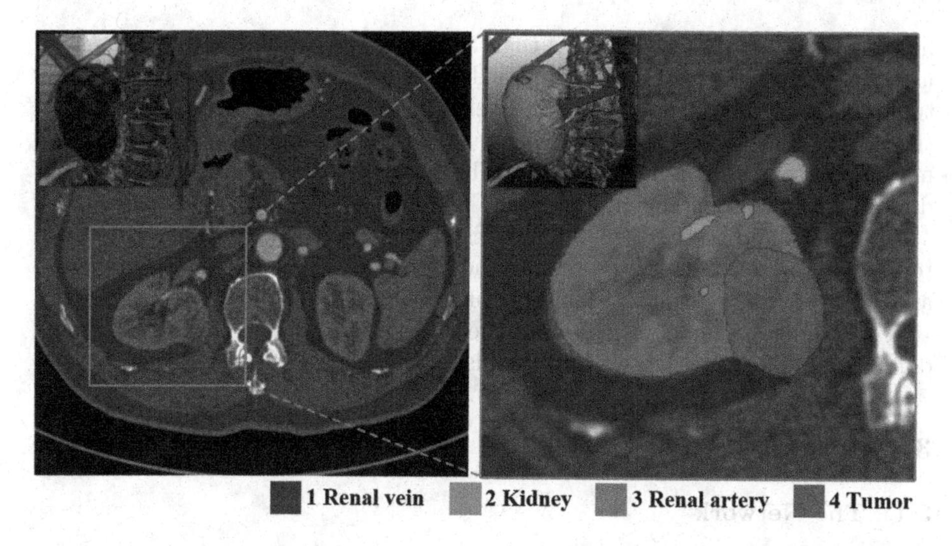

Fig. 1. Preview of a data sample as shown in KiPA 2022 [1] dataset description.

artery, and renal vein. These structures present challenges in shape variation, distribution variation, class imbalance, and low significance. The KiPA 2022 challenge has collected 130 images for training and testing and uses Dice, Hausdorff Distance (HD), and Average Hausdorff Distance (AVD) as evaluation metrics. This challenge is expected to promote renal cancer treatment, research interactions, and interdisciplinary communication. Additionally, the challenge aims to encourage the development of new algorithms and techniques for segmenting these structures in CTA images and to foster collaboration and exchange of ideas among researchers in the field.

In this work, we describe our approach for 3D segmentation of renal vein, kidney, renal artery, and tumor from KiPA 2022 dataset. Our method achieved 10th position in the open test leaderboard and 6th position in the closed test leaderboard.

2 Dataset

The KiPA 2022 dataset [1] consists of unenhanced abdominal CT images from 130 patients. The images were annotated by mature algorithms and then adjusted by 3 experts who received training on annotation. The images were acquired using a Siemens dual-source 64-slice CT scanner, and the kidney tumor types consist of clear renal cell carcinomas, papillary, chromophobe, angiomyolipoma, and eosinophilic adenoma. The sizes of the kidneys and tumors in the images range from 73.73 ml to 263.02 ml and 2.06 ml to 646.68 ml, respectively.

The preprocessing steps applied to the images include resampling so that the resolution of the z-axis is the same as that of the x-/y-axis and expanding the kidney and tumor labels to a maximum of 32 pixels to crop the regions of interest (ROI). If the image is less than 32 pixels away from the kidney region, it is cropped at the maximum amplified pixel distance to include the kidney, tumor, part of the renal arteries, and veins in the ROI. Privacy information is then removed, and the remaining image content and original resolution information are saved in the nii.gz format. The subtypes of renal tumors are evenly distributed in training, open, and closed testing sets.

In the ground-truth voxel value label files, 0 represents the background, 1 represents the renal vein, 2 represents the kidney, 3 represents the renal artery, and 4 represents the tumor. Figure 1 visualizes a data sample.

For training our models, we used the KiPA 2022 dataset [1] only. We randomly split the entire dataset into 5-folds and trained a model for each.

3 Methods

3.1 The Network

We implemented our approach with MONAI[1] [2]. We use the encoder-decoder backbone based on [5] with an asymmetrically larger encoder to extract image features and a smaller decoder to reconstruct the segmentation mask [6–8]. The network is summarized in Table 1.

Encoder Part. The encoder part of the model uses ResNet [3] blocks a type of convolutional neural network (CNN) designed to improve the training of deep networks. The ResNet architecture introduces an identity skip connection, which allows the gradient to flow directly from the output of a layer to its input, improving the flow of information through the network and making it easier to train.

We have used 5 stages of down-sampling in the encoder, each containing a different number of convolutional blocks: 1, 2, 2, 4, and 4, respectively. We have used batch normalization [4] and ReLU activation functions in the convolutional blocks, which help improve the model's performance and stability.

The output of each block is followed by an additive identity skip connection, which allows the gradient to flow directly from the output of the block to its input. This helps improve the information flow through the network and makes training easier.

We follow a common CNN approach to progressively downsize the image dimensions by a factor of 2 and simultaneously increase the feature size by 2. This allows us to reduce the spatial resolution of the input images while increasing the number of features extracted from each image. For downsizing, we use strided convolutions, which allow us to reduce the spatial dimensions of the input without increasing the number of parameters in the model.

All convolutions in the encoder are $3 \times 3 \times 33$ with an initial number of filters equal to 32. The encoder is trained with $128 \times 128 \times 3160$ input regions, which

[1] https://github.com/Project-MONAI/MONAI.

means it processes images with dimensions $128 \times 128x160$. This relatively small input size allows us to train the model more efficiently but may also limit its ability to capture fine-grained details in the images.

Decoder Part. The decoder structure is similar to the encoder structure but with a single block per each spatial level. This means that the decoder has the same number of stages as the encoder but only one block in each stage. This allows the decoder to gradually recover the spatial resolution of the input images while maintaining many features.

Each decoder level begins with upsizing using a transposed convolution, which reduces the number of features by a factor of 2 and doubles the spatial dimension. This allows the decoder to recover the spatial resolution of the input images while maintaining many features. The output of the upsampling operation is then added to the encoder output of the equivalent spatial level, which helps to propagate information from the encoder to the decoder and improve the model's performance.

The end of the decoder has the same spatial size as the original image, and the number of features equals the initial input feature size. This means that the decoder recovers the spatial resolution of the input images while maintaining the original number of features. The output of the decoder is then passed through a $1 \times 1x1$ convolution to produce 8 channels, followed by a softmax activation function. This allows the model to produce a probability map for each of the 8 classes in the input images.

3.2 Data Preprocessing and Augmentation

We normalize all input images to have zero mean and unit standard deviation (based on nonzero voxels only). This means that we subtract the mean value of the nonzero voxels from each image and divide it by the standard deviation of the nonzero voxels. This normalization step is commonly used to improve the performance of deep learning models by ensuring that the input data has a consistent scale and distribution.

We have applied several data augmentation techniques to the input images to increase the training data's diversity and improve the model's generalization. These techniques include random rotation, random zoom on each axis, random Gaussian smoothing, and random Gaussian noise with a probability of 0.2. We have also applied a random flip on each axis and random contrast adjustment with a probability of 0.5. These augmentation techniques help to prevent overfitting by introducing variability and noise into the training data, which forces the model to learn more robust and generalizable features.

3.3 Training Details

We have used Dice Focal loss for training our model. This loss function is a variant of the Dice loss, commonly used in medical image segmentation tasks,

Table 1. Backbone network structure, where IN stands for group normalization, Conv - $3 \times 3 \times 33$ convolution, AddId - addition of identity/skip connection.

Name	Ops	Repeat
InitConv	Conv	1
EncoderBlock0	IN, ReLU, Conv, IN, ReLU, Conv, AddId	1
EncoderDown1	Conv stride 2	1
EncoderBlock1	IN, ReLU, Conv, IN, ReLU, Conv, AddId	2
EncoderDown2	Conv stride 2	1
EncoderBlock2	IN, ReLU, Conv, IN, ReLU, Conv, AddId	2
EncoderDown3	Conv stride 2	1
EncoderBlock3	IN, ReLU, Conv, IN, ReLU, Conv, AddId	4
EncoderDown4	Conv stride 2	1
EncoderBlock4	IN, ReLU, Conv, IN, ReLU, Conv, AddId	4
DecoderUp3	UpConv, +EncoderBlock3	1
DecoderBlock3	IN, ReLU, Conv, IN, ReLU, Conv, AddId	1
DecoderUp2	UpConv, +EncoderBlock2	1
DecoderBlock2	IN, ReLU, Conv, IN, ReLU, Conv, AddId	1
DecoderUp1	UpConv, +EncoderBlock1	1
DecoderBlock1	IN, ReLU, Conv, IN, ReLU, Conv, AddId	1
DecoderUp0	UpConv, +EncoderBlock0	1
DecoderBlock0	IN, ReLU, Conv, IN, ReLU, Conv, AddId	1
DecoderEnd	IN, ReLU, Conv1, Sigmoid	1

and incorporates a modulating factor (the focal loss) to down-weight the loss assigned to well-classified samples.

We use the AdamW optimizer with an initial learning rate of $2e^{-4}$ and decrease it to zero at the end of the final epoch using the Cosine annealing scheduler. This allows us to adapt the learning rate during training to improve the convergence of the model. We have used a batch size of 8, which means that the model processes 8 images at a time during each iteration of training. The model is trained using 8 GPUs, with each GPU optimizing for a batch size of 1. However, we have calculated batch normalization across all the GPUs to improve the model's performance.

We have ensembled 10 models for final submission to the KiPA 2022 challenge. This means that we have trained 10 individual models and combined their predictions by averaging them to produce the final result. All the models were trained for 1000 epochs.

We use L2 norm regularization on the convolutional kernel parameters with a weight of $1e^{-5}$. This regularization technique helps to prevent overfitting by adding a penalty term to the loss function that encourages the model to learn smaller weights. The weight of $1e^{-5}$ controls the strength of the regularization, with a larger weight leading to smaller weights in the model.

4 Results

Table 2 shows the average DICE, HD, and AVD scores for different structures in the open and closed test sets of the KiPA 2022 challenge. The table shows that the kidney structure has the highest average DICE score in both the open and closed test sets, followed by the renal artery, tumor, and renal vein structures. The HD and AVD scores are also shown, with the kidney and renal artery structures having the lowest average HD and AVD scores in both test sets. These results indicate that the algorithms used in the KiPA 2022 challenge were able to segment the kidney and renal artery structures with high accuracy. In contrast, the tumor and renal vein structures were more difficult to segment. Our method achieved 10th position in the open test set and 6th position in the closed test set.

Table 2. Summary of our method's result on the open and closed test set of KiPA 2022 challenge. Our method achieved 10th position in the open test set and 6th position in the closed test set.

Target structure	Open test set			Closed test set		
	Dice	HD	AVD	Dice	HD	AVD
Vein	0.8197	12.7067	0.9175	0.8275	12.6844	0.6544
Kidney	0.9561	17.0958	0.4743	0.9583	17.3373	0.4621
Artery	0.8965	19.4163	0.3603	0.9003	14.1877	0.2388
Tumor	0.8708	13.3774	1.6256	0.8900	11.6567	1.0998

5 Conclusion

In conclusion, our approach for 3D segmentation of renal vein, kidney, renal artery, and tumor in the KiPA 2022 challenge achieved competitive performance, achieving 10th position in the open test leaderboard and 6th position in the closed test leaderboard. Our method, based on the MONAI framework and the SegResNet architecture, demonstrates the effectiveness of deep learning for the automated segmentation of renal structures in CTA images. This approach can potentially improve the accuracy and reliability of renal cancer treatment, providing detailed and high-resolution information about the structures of interest. We believe our approach can be further improved and extended to other medical imaging tasks, and we look forward to future developments in the field.

References

1. Kipa 2022. https://kipa22.grand-challenge.org/dataset/
2. Project-monai/monai. https://doi.org/10.5281/zenodo.5083813
3. He, K., Zhang, X., Ren, S., Sun, J.: Deep residual learning for image recognition. In: Proceedings of the IEEE Conference on Computer Vision and Pattern Recognition, pp. 770–778 (2016)

4. Ioffe, S., Szegedy, C.: Batch normalization: accelerating deep network training by reducing internal covariate shift. In: International Conference on Machine Learning, pp. 448–456. PMLR (2015)
5. Myronenko, A.: 3D MRI brain tumor segmentation using autoencoder regularization. In: Crimi, A., Bakas, S., Kuijf, H., Keyvan, F., Reyes, M., van Walsum, T. (eds.) BrainLes 2018. LNCS, vol. 11384, pp. 311–320. Springer, Cham (2019). https://doi.org/10.1007/978-3-030-11726-9_28
6. Ronneberger, O., Fischer, P., Brox, T.: U-Net: convolutional networks for biomedical image segmentation. In: Navab, N., Hornegger, J., Wells, W.M., Frangi, A.F. (eds.) MICCAI 2015. LNCS, vol. 9351, pp. 234–241. Springer, Cham (2015). https://doi.org/10.1007/978-3-319-24574-4_28
7. Zhou, Z., Rahman Siddiquee, M.M., Tajbakhsh, N., Liang, J.: UNet++: a nested U-Net architecture for medical image segmentation. In: Stoyanov, D., et al. (eds.) DLMIA/ML-CDS -2018. LNCS, vol. 11045, pp. 3–11. Springer, Cham (2018). https://doi.org/10.1007/978-3-030-00889-5_1
8. Zhou, Z., Siddiquee, M.M.R., Tajbakhsh, N., Liang, J.: Unet++: redesigning skip connections to exploit multiscale features in image segmentation. IEEE Trans. Med. Imaging **39**(6), 1856–1867 (2019)

Ensembled Autoencoder Regularization for Multi-structure Segmentation for Kidney Cancer Treatment

David Jozef Hresko[1(✉)], Marek Kurej[1], Jakub Gazda[2], and Peter Drotar[1]

[1] IISlab, Technical University of Kosice, Kosice, Slovakia
david.jozef.hresko@tuke.sk

[2] 2nd Department of Internal Medicine, Pavol Jozef Safarik University and Louis Pasteur University Hospital, Kosice, Slovakia

Abstract. The kidney cancer is one of the most common cancer types, which treatment frequently include surgical intervention. However, surgery is in this case particularly challenging due to regional anatomical relations. Organ delineation can significantly improve surgical planning and execution, especially in an automated way that saves time and does not require skilled clinicians. In this contribution, we propose ensemble of two fully convolutional networks for segmentation of kidney, tumor, veins and arteries. While SegResNet architecture achieved better performance on tumor, the nnU-Net provided more precise segmentation for kidneys, arteries and veins. So in our proposed approach we combine these two networks, and further boost the performance by mixup augmentation. With mentioned approach we achieved 10th place in the KiPA2022 challenge.

Keywords: Segmentation · Kidney · SegResNet · nnU-Net · Ensemble · Mixup

1 Introduction

Kidney cancer (KC) has become one of the ten most common cancer types in the general population, while its incidence steadily increased from the 1970 s s until the mid-1990s. This trend was attributed to the improved diagnosis secured by the broad introduction of advanced radiological imaging to clinical practice [1]. However, KC is strongly associated with risk factors such as cigarette smoking, obesity, and arterial hypertension. Therefore, even though the incidence has recently levelled off, it is not expected to drop anytime soon. On the contrary, it will pose a significant threat in industrialized countries. The treatment of KC is based on ablation, chemotherapy, and surgery [2,11]. Luckily, most KCs are detected early and only incidentally by radiologic imaging for other diseases when the curative treatment - surgery - is still possible. However, surgery is specifically challenging due to complex regional anatomical relations. Kidney parsing

Y. Xiao et al. (Eds.): CuRIOUS/KiPA/MELA 2022, LNCS 13648, pp. 43–51, 2023.
https://doi.org/10.1007/978-3-031-27324-7_6

can help tackle this issue and improve preoperative planning and perioperative decisions, leading to a higher chance of successful tumour resection.

Abdominal organ segmentation from medical imaging is a topic that has been researched for several years. Particularly, the computed tomography can provide detailed information about organs and structures in abdominal area. This information can be used for surgical planning, diagnosis and various clinical decisions. In recent years the contour delineation of anatomical and pathological structures became important for applications such as visual support during surgery. Since the manual delineation is tedious and time consuming the automated highly accurate methods are strongly desired.

The advent of convolutional neural networks brought many efficient solution also for medical images segmentation. The fully convolutional neural network, particularly U-Net architecture [9] became de-facto golden standard for medical image segmentation. Moreover, the pipeline optimisation process implemented in nnU-Net [5] boosted segmentation performance and made the U-net easily applicable for any segmentation task. Considering all these recent advances, the abdominal organ segmentation appears to be almost solved problem [6]. However, even though this may be partially true for a big organs such as liver, kidneys and spleen, the smaller structures still represent challenge for automated segmentation approaches.

Challenges, like LiTS, KiTS, FLARE did not consider smaller structure namely arteries and veins. However segmentation of arteries and veins is crucial for urological surgeries namely laparoscopic nephrectomy [8,10]. In this case the precise information about target segmental arteries is required in order to avoid insufficient clamping.

In this paper we propose ensemble of two encoder-decoder based convolutional neural networks: nnU-net and SegResNet [7]. To further boost the performance of nnU-net, we additionally applied manifold mixup augumentation [12]. Similar improvement was already proposed by authors of [3] on KiTS21 challenge, which improved overall performance of their solution for kidney and kidney tumor segmentation task.

The rest of the paper is organized as follows. In the next section we provide detailed data description. In the methodology section, the proposed solution is explained. Finally we present the results and discuss different aspects of our submission.

2 Data

The data were acquired by Siemens dual-source 64-slice CT scanner and the contrast media was injected during CT image acquisition. The further technical settings of CT are: X-ray tube current is 480 mA, B25F convolution kernel, exposure time equal to 500 ms and voltage 120KV. The slice thickness is 0.5mm/pixel and spacing of images is from 0.47 mm/pixel to 0.74 mm/pixel and 0.75 mm/pixel for z-direction, respectively.

Altogether there are 130 3D abdominal CT images. The ground-truth corresponding to four classes, kidney, tumor, vein and artery, was determined by three medically trained experts and validated by experienced radiologist. All images were cropped to the same size of $150 \times 150 \times 200$ to focus on the four structures of interests

From 130 cases, 70 are used for training and validation of the model, 30 for open test phase evaluation and 30 for closed test phase evaluation.

3 Methodology

Besides the state-of-art architecture nnU-Net, which is commonly used to solve medical segmentation tasks, we decided to also adapt and verify not so well-known architectures. In particular, we chose autoencoder based architecture with additional regularization named SegResNet. Based on the obtained results from training phase we decided to ensemble the output from nnU-Net and SegResNet, which were later additionally fine-tuned for closed test phase of KiPA22 challenge.

3.1 Preprocessing

In case of the nnU-Net architecture, preprocessing methods including transformation, re-sampling, normalization and scaling were automatically handled by the nnU-Net pipeline. This includes, the resampling strategy for anisotropic data was third order spline interpolation for in-plane and nearest neighbours for out-of-plane. Global dataset percentile clipping along with z-score with global foreground mean and standard deviation was chosen as a normalization strategy. The clipping of HU values were automatically preset with default setting (0.5 and 99.5 percentile).

In case of the SegResNet architecture, as the first step we applied regular channel-wise normalization and scaling to clip CT scan values. After that, several augmentations were applied, each with the probability of 0.5. We randomly added gaussian noise to original scan, followed by random rotation, zooming, axis flipping and elastic deformations using bilinear interpolation to calculate output values.

3.2 Post-processing

Based on our observations during training phase, we noticed that some faulty segmentations of kidney and tumor structures were mainly caused by small redundant segments, which were located far from the expected structure location. To prevent this phenomena, we decided to apply keep largest connected component method as the post-processing step during separate inference of both models.

3.3 Proposed Approach

Our proposed method relies on ensemble technique and consists of two different encoder-decoder based variations of U-Net architecture. The first one is the nnU-Net architecture, which already dominated several segmentation challenges. Following the approach of [3] we also took advantage of mixup augmentation. Instead of regular version of mixup [13], we utilized extended version named manifold mixup [12]. The key difference here is that, the mixing process of batch samples can be also applied on inner layers of the network, not only on the input layer. The mixed minibatch is then passed forward regularly from k-th inner layer to the output layer, which is used to compute the loss value and gradients.

During extensive experiments performed on KiPA22 public dataset we discovered that, there is an architecture, which is capable of better results on some of the segmented classes. Concretely, performance of the nnU-Net was outperformed by SegResNet on tumor class. To benefit from both architectures, we decided to perform ensemble of these two architectures to better localize targeted structures. Based on the obtained results from individual architectures, we keep predictions of kidney, renal vein and renal artery class from nnU-Net, while tumor were produced by SegResNet. Overall concept of proposed method is depicted in Fig. 1.

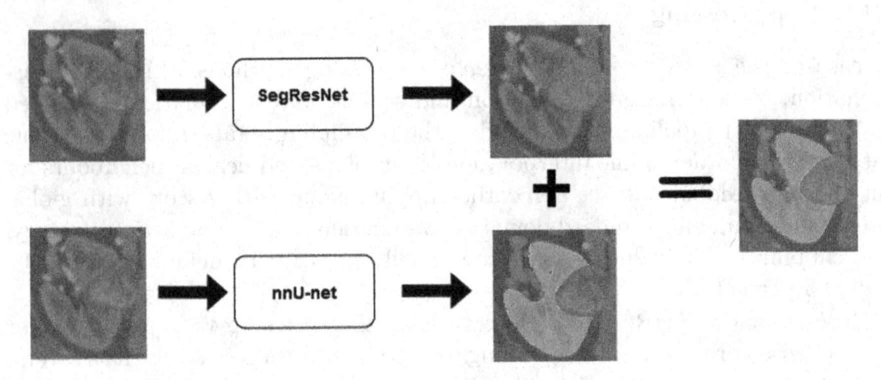

Fig. 1. Overview of proposed ensemble method

4 Results

The evaluation of the model performance is based on two different metrics. The first one relies on the area based metric Dice Similarity Coefficient (DSC). DSC is used to evaluate the area-based overlap index. The second one is based on distance between predicted and ground truth segmentation mask. Here, to evaluate the coincidence of the surface for less sensitive to outliers, Average Hausdorff Distance (AHD) is used. Additionally, outliers sensitive Hausdorff Distance (HD) is also used to further evaluate the segmentation quality.

To train the final version of nnU-Net, we used stochastic gradient descent (SGD) optimizer with the initial learning rate of 0.01. The length of training was set to 1000 epochs. For manifold mixup augmentation we set the hyperparameter value of mixing coefficient to $\alpha = 0.1$. In case of the final version of SegResNet, we used AdamW optimizer with the initial learning rate of 0.0001. The length of training was set to 4000 epochs. For both models we used combined Dice and Cross Entropy loss, batch size equal to two.

The results from the training phase, along with comparisons to other experiments with different models and configurations are presented in Table 1. As can be seen, proposed manifold mixup improved performance of the nnU-Net for each class. SegResNet outperformed nnU-Net for tumor class segmentation task with notable differences. Based on the [4] we were also curious if transformer based architecture is truly capable of better results than standard convolutional network, so we also trained UNETR model on this dataset. Our results proved that this vision transformer was not able to reach the performance of any baseline version of tested models.

Table 1. Overall performance of examined models on KiPA22 train data

	Kidney			Tumor			Vein			Artery		
Network	DSC	HD	AHD	DSC	HD	AHD	DSC	HD	AHD	DSC	HD	AHD
nnU-Net (baseline)	0.960	17.108	0.484	0.856	12.432	1.412	0.805	12.002	1.180	0.839	15.995	0.512
nnU-Net + mixup	**0.963**	**17.022**	**0.424**	0.893	10.114	1.225	**0.823**	12.112	**0.830**	0.849	**16.395**	**0.449**
UNETR	0.951	19.434	0.874	0.838	14.768	1.992	0.773	18.109	3.556	0.822	20.542	1.987
SegResNet	0.961	17.045	0.512	**0.901**	**9.100**	**0.970**	0.818	**11.224**	1.952	**0.873**	19.156	2.443

The Fig. 2 shows the example prediction of our proposed method. We randomly selected case 61 from public dataset, which contains all four labels. As can be seen, the segmentation of all structures is almost identical with the ground

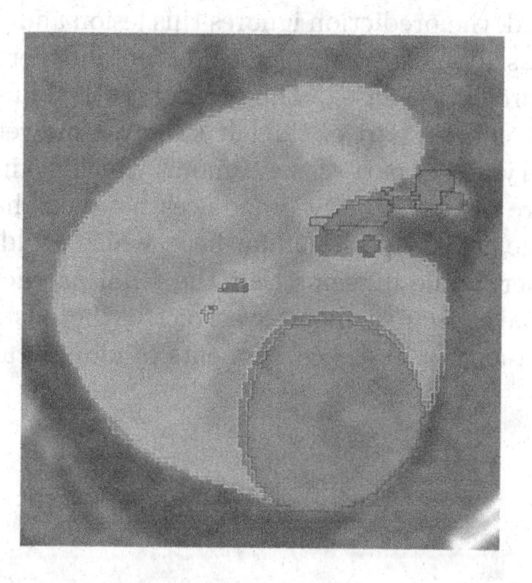

Fig. 2. Example segmentation for the 61 case. Individual contours denote ground truth label for specific class. Predicted segmentation classes are: yellow = kidney, blue = tumor, green = renal vein, red = renal artery. (Color figure online)

Table 2. Overall performance of proposed method on KiPA22 test data

	Kidney			Tumor			Vein			Artery		
Phase	DSC	HD	AHD	DSC	HD	AHD	DSC	HD	AHD	DSC	HD	AHD
Open test	0.957	17.008	0.464	0.880	9.104	1.616	0.835	12.733	0.322	0.878	16.195	0.322
Closed test	0.957	19.766	0.627	0.865	11.123	1.630	0.878	17.641	0.515	0.831	14.201	0.669

truth regions. The yellow colour represents healthy kidney tissue located laterally to renal vessels. The red and green colours medially represent the renal artery and renal vein, respectively. Finally, the blue colour dorsally represents kidney cancer. This picture informs the surgeon that the kidney cancer has not invaded renal vessels yet and has a high chance of successful resection. However, a surgeon must evaluate all sections, which may present different spatial relation between kidney cancer and renal vessels. In more complicated cases, human visual evaluation might not be enough to evaluate the ground truth correctly.

Overall performance of our proposed method was additionally, independently measured on KiPA2022 challenge data during open and closed test phase of the challenge. The obtained results can be seen in the Table 2.

5 Discussion

Here we investigate in detail some of the segmentation cases that reached lower scores. Figure 3a represents "a double incorrect case" because neither the ground truth nor the prediction is correct. First, the ground truth (yellow contour, representing healthy renal parenchyma) contains a pathological (probably cystic) lesion (it is impossible to tell without considering other phases of the contrast enhancement). Second, the prediction ignores this lesion and goes around it without rendering and classifying the pathological lesion. Furthermore, Fig. 3b represents "an incorrect prediction case". The ground truth (blue contour) depicts a tumour located in the dorsal part of the left kidney. However, the segmentation fails to render the very dorsal part of the tumour, which again appears as its cystic component. There are no density differences between the cystic component of the tumour and the healthy perirenal fat here, which could be why the prediction failed. Altogether, while it seems that the renal parenchyma, renal vessels and solid tumours have been segmented correctly, the mistakes are associated mostly with cystic lesions or cystic components of kidney cancers.

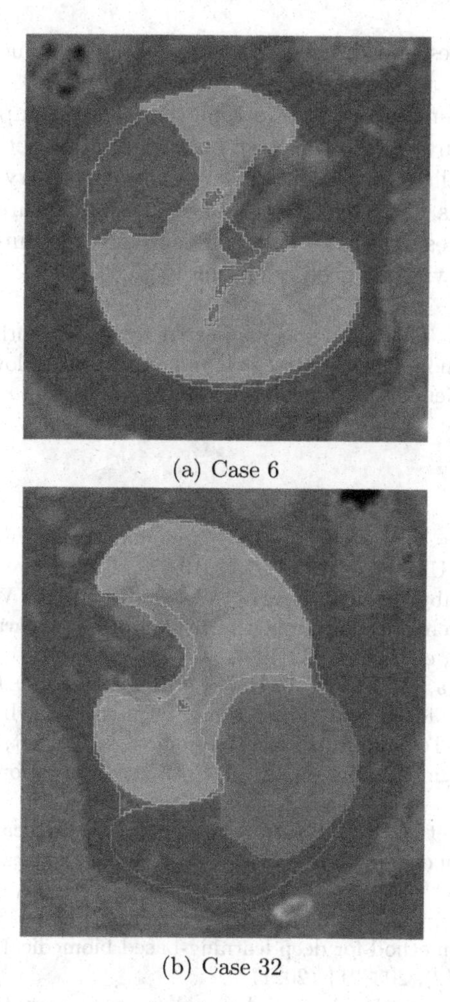

(a) Case 6

(b) Case 32

Fig. 3. Examples of faulty segmentations. Individual contours denote ground truth label for specific class. Predicted segmentation classes are: yellow = kidney, green = renal vein, red = renal artery. (Color figure online)

6 Conclusions

In this paper we proposed ensemble method for segmentation of structures in abdominal area. The ensemble is build on the combination of nnU-Net and SegResNet architecture. The proposed approach minimizes the probability of over-fitting by using additional mixup augmentation.

During the experimentation phase wc tested multiple models with various improvements and performed tuning of the hyperparameters to find the optimal solution. However, member of our team, who is skilled clinician, performed manual analysis of model's performance by evaluating the segmentation predictions on some training cases and discovered weaknesses of our model, which need to be addressed in case of production application.

With mentioned enhancements we achieved highly competitive score on all four targeted structures and 10th place on the test dataset of KiPA2022 challenge. For future challenges and our research we plan to try other variations of mixup augmentations not only on the nnU-Net architecture. One of the other possible way of the research is to enhance SegResNet architecture to overcome nnU-Net on its own without need of ensemble method.

Acknowledgements. This work was supported by the Scientific Grant Agency of the Ministry of Education, Science, Research and Sport of the Slovak Republic and the Slovak Academy of Sciences under contract VEGA 1/0327/20.

References

1. Chow, W.H., Dong, L.M., Devesa, S.S.: Epidemiology and risk factors for kidney cancer. Nat. Rev. Urol. **7**(5), 245–257 (2010)
2. Dahle, D.O., Skauby, M., Langberg, C.W., Brabrand, K., Wessel, N., Midtvedt, K.: Renal cell carcinoma and kidney transplantation: a narrative review. Transplantation **106**(1), e52–e63 (2022)
3. Gazda, M., Bugata, P., Gazda, J., Hubacek, D., Hresko, D.J., Drotar, P.: Mixup augmentation for kidney and kidney tumor segmentation. In: Heller, N., Isensee, F., Trofimova, D., Tejpaul, R., Papanikolopoulos, N., Weight, C. (eds.) Kidney and Kidney Tumor Segmentation, pp. 90–97. Springer International Publishing, Cham (2022)
4. Hatamizadeh, A., et al.: UNETR: transformers for 3D medical image segmentation (2021). https://doi.org/10.48550/ARXIV.2103.10504. https://arxiv.org/abs/2103.10504
5. Isensee, F., Jaeger, P.F., Kohl, S.A.A., Petersen, J., Maier-Hein, K.H.: nnU-Net: a self-configuring method for deep learning-based biomedical image segmentation. Nat. Methods **18**(2), 203–211 (2021)
6. Ma, J., et al.: AbdomenCT-1K: Is abdominal organ segmentation a solved problem? (2020). https://doi.org/10.48550/ARXIV.2010.14808. https://arxiv.org/abs/2010.14808
7. Myronenko, A.: 3D MRI brain tumor segmentation using autoencoder regularization (2018). https://doi.org/10.48550/ARXIV.1810.11654. https://arxiv.org/abs/1810.11654
8. Porpiglia, F., Fiori, C., Checcucci, E., Amparore, D., Bertolo, R.: Hyperaccuracy three-dimensional reconstruction is able to maximize the efficacy of selective clamping during robot-assisted partial nephrectomy for complex renal masses. Eur. Urol. **74**(5), 651–660 (2018)
9. Ronneberger, O., Fischer, P., Brox, T.: U-Net: convolutional networks for biomedical image segmentation. In: Navab, N., Hornegger, J., Wells, W.M., Frangi, A.F. (eds.) MICCAI 2015. LNCS, vol. 9351, pp. 234–241. Springer, Cham (2015). https://doi.org/10.1007/978-3-319-24574-4_28
10. Shao, P., et al.: Laparoscopic partial nephrectomy with segmental renal artery clamping: technique and clinical outcomes. Eur. Urol. **59**(5), 849–855 (2011)

11. Shao, P., et al.: Precise segmental renal artery clamping under the guidance of dual-source computed tomography angiography during laparoscopic partial nephrectomy. Eur. Urol. **62**(6), 1001–1008 (2012)
12. Verma, V., et al.: Manifold mixup: better representations by interpolating hidden states (2018). https://doi.org/10.48550/ARXIV.1806.05236. https://arxiv.org/abs/1806.05236
13. Zhang, H., Cisse, M., Dauphin, Y.N., Lopez-Paz, D.: mixup: beyond empirical risk minimization (2017). https://doi.org/10.48550/ARXIV.1710.09412. https://arxiv.org/abs/1710.09412

The 2022 Correction of Brain Shift with Intra-Operative Ultrasound-Segmentation Challenge (CuRIOUS-SEG 2022)

Segmentation of Intra-operative Ultrasound Using Self-supervised Learning Based 3D-ResUnet Model with Deep Supervision

Abdul Qayyum[1,4](✉), Moona Mazher[2], Steven Niederer[3], and Imran Razzak[4]

[1] ENIB, UMR CNRS 6285 LabSTICC, 29238 Brest, France
qayyum@enib.fr
[2] Department of Computer Engineering and Mathematics, University Rovira I Virgili, Tarragona, Spain
[3] Department of Biomedical Engineering, King's College London, London, UK
Steven.niederer@kcl.ac.uk
[4] School of Computer Science and Engineering, University of New South Wales, Sydney, Australia
imran.razzak@ieee.org

Abstract. Intra-operative ultrasound (iUS) is a robust and relatively inexpensive technique to track intra-operative tissue shift and surgical tools. Automatic algorithms for brain tissue segmentation in iUS, especially brain tumors and resection cavity can greatly facilitate the robustness and accuracy of brain shift correction through image registration, and allow easy interpretation of the iUS. This has the potential to improve surgical outcomes and patient survival rates. In this paper, we have proposed a self-supervised two-stage model for the Intra-operative ultrasound (iUS) task. In the first stage, we trained the encoder of our proposed 3DResUNet model using the self-supervised contrastive learning. The self-supervised learning offers the promise of utilizing unlabeled data. The training samples are used in self-supervision to train the encoder of the proposed 3DResUNet model and utilized this encoder as a pre-trained weight for the Intra-operative ultrasound (iUS) segmentation. In the second stage, the pre-trained weighted-based 3DResUNet proposed model was used to train on the training dataset for iUS segmentation. Experiment on CuRIOUS -22 challenge showed that our proposed solution showed significantly better performance before, during, and after Intra-operative ultrasound (iUS) segmentation. The code is publicly available (https://github.com/RespectKnowledge/SSResUNet_Intra-operative-ultrasound-iUS-Tumor-Segmentation).

1 Introduction

Intra-operative ultrasound is a high-energy sound waves surgical procedures that are bounced off internal tissues and organs. It is a low-cost dynamic imaging modality that provides provides interactive and timely information during surgery which helps the surgeon to find tumours or other problems during the procedure. As the transducer is in direct contact with the organ being examined, hence, it can guide the surface incisions for deep lesion resection, limit the extent of surgical resection, accurately

Y. Xiao et al. (Eds.): CuRIOUS/KiPA/MELA 2022, LNCS 13648, pp. 55–62, 2023.
https://doi.org/10.1007/978-3-031-27324-7_7

guide intraoperative biopsies, localize the pathology and we can obtain high-resolution images which are not degraded by air, bone, or overlying soft tissues.

For most cancers, the survival at one and five years is much higher if it is detected at early (stage 1) than later stage. When a cancer diagnosis is delayed or inaccessible, the survival chances decrease significantly, and it may have greater problems associated with treatment and much higher costs of care i.e., 90% of patients has 10+ year survival rate after being diagnosed in early stages in comparison to 5% for those who are diagnosed at later stage (stage 4).

Automatic algorithms for brain tissue segmentation in iUS, especially brain tumors and resection cavity can greatly facilitate the robustness and accuracy of brain shift correction through image registration and allow easy interpretation of the iUS. The resection quality and safety are often affected by the intra-operative brain tissue shift due to several factors i.e. intracranial pressure change, drug administration, gravity and tissue removal. Such shift in tissue may results in displacement of target and vital structures during surgical procedure while the displacements may not be directly visible in the surgeon however, it renders the surgical plan invalid. Live ultrasound overlaid onto pre-operative data which allows for assessment and visualization of brain shift. The images from intra-operative ultrasound contain biological information possibly correlated to the tumour's behaviour, aggressiveness, and oncological outcomes. Deep Learning has been widely used in overall medical image segmentation tasks [1–5].

To tackle the discrepancies of iUS at different surgical stages, in this work, we present an efficient segmentation of intra-operative ultrasound using self-supervised learning-based 3D-ResUnet with deep supervision. We trained the encoder of our proposed 3DResUNet model using the self-supervised contrastive learning method. The self-supervised learning offers the promise of utilizing unlabeled data. The training sample is used in self-supervision to train the encoder of the proposed 3DResUNet model and utilized this encoder as a pre-trained weight for the Intra-operative ultrasound (iUS) segmentation task. In the second stage, the pre-trained weighted-based 3DResUNet proposed model was used to train on the training dataset for iUS segmentation. Our proposed solution produced optimal performance on the validation dataset for three tasks (before, during, and after Intra-operative ultrasound (iUS) segmentation).

2 Methods

In self-supervised learning sitting, First, it uses augmentation to mutate the data, and second, it utilizes regularized contrastive loss [6] to learn feature representations of the unlabelled data. The multiple augmentations are applied on a randomly selected 3D foreground patch from a 3D volume. Two augmented views of the same 3D patch are generated for the contrastive loss as it functions by drawing the two augmented views closer to each other if the views are generated from the same patch, if not then it tries to maximize the disagreement. We have to use masked volume inpainting, contrastive learning, and rotation prediction as proxy tasks for learning contextual representations of input images. The primary task of the network is to reconstruct the original image. The different augmentations used are classical techniques such as in-painting [7], out-painting, and noise augmentation to the image by local pixel shuffling [8]. The secondary task of the network is to simultaneously reconstruct the two augmented views as similar to each other as possible via regularized contrastive loss [6] as its objective is to maximize the agreement. The term regularized has been used here because contrastive loss is adjusted by the reconstruction loss as a dynamic weight itself. Multiple patches having sizes 128 × 128 × 128 are generated and used different views based on the augmentation via the transforms on the same cubic patch. The objective of the SSL network is to reconstruct the original image. The contrastive loss is driven by maximizing agreement of the reconstruction based on the input of the two augmented views.

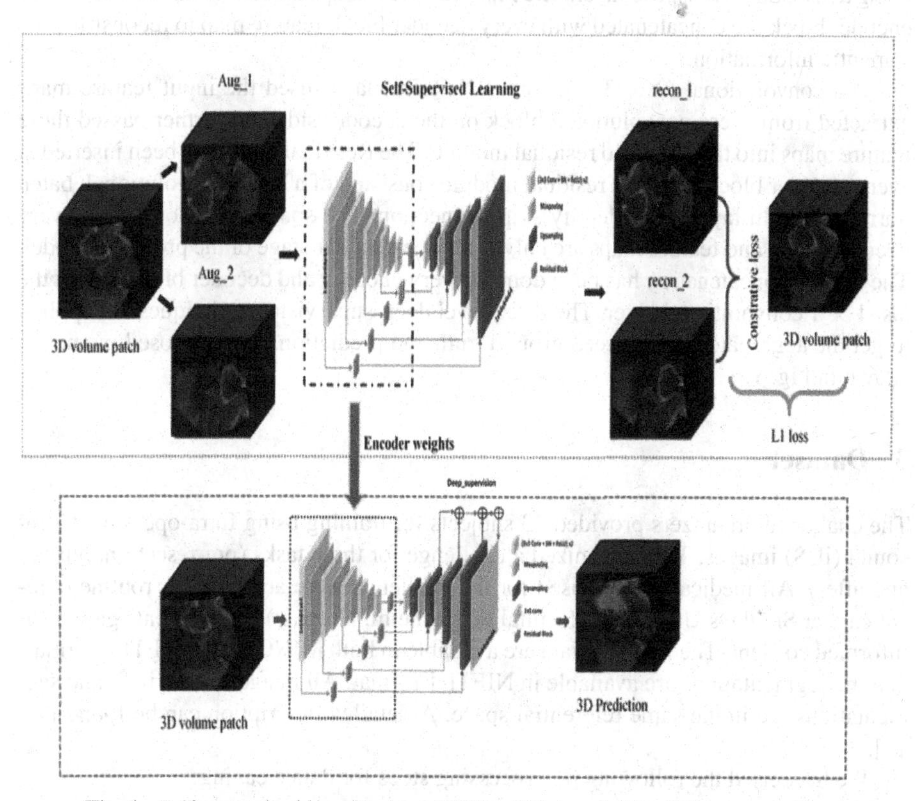

Fig. 1. Self-supervised learning based 3DResUNet model for iUS segmentation

The pre-trained encoder weights are used in the proposed 3DResUNet with deep supervision before, during, and after the iUS segmentation task. We trained our proposed model using a training supervised dataset based on patches. We randomly generated patches with size 128 × 128 × 128 from input volume and used different augmentations to train the proposed model. We used a sliding window with 128 strides to generate the prediction volume on the validation dataset. The training transforms such as RandCrop, RandGaussianNoise, RandGaussianSmooth, RandShiftIntensity, RandAdjustContrast, and RandZoomd were used to train the proposed 3DResUNet model. All transformations were used from MONAI Library [9].

3D-ResUnet with Deep Supervision: A framework of the proposed model is presented as an encoder, a decoder, and a baseline module. The 1×1 convolutional layer with softmax function has been used at the end of the proposed model. The 3D strides convolutional layer has been used to reduce the input image spatial size. The convolutional block consists of convolutional layers with Batch-Normalization and ReLU activation function to extract the different feature maps from each block on the encoder side. In the encoder block, the spatial input size has been reduced with an increasing number of feature maps and on the decoder side, the input image spatial size will increase using a 3D Conv-Transpose layer. The input features' maps that are obtained from every encoder block are concatenated with every decoder block feature map to reconstruct the semantic information.

The convolutional ($3 \times 3 \times 3$conv-BN-ReLu) layer used the input feature maps extracted from every convolutional block on the encoder side and further passed these feature maps into the proposed residual module. The Residual block has been inserted at every encoder block. Each 3d residual module consisted of a 3×3 convolutional, batch norm, and Relu layer with identity skip connection. The spatial size doubled at every decoder block and feature maps are halved at each decoder stage of the proposed model. The feature concatenation has been done at every encoder and decoder block except the last 1×1 convolutional layer. The three-level deep-supervision techniques are applied to get the aggregate loss between ground truth and prediction. The proposed method is shown in Fig. 1.

3 Dataset

The challenge organizers provided 23 subjects for training using Intra-operative Ultrasound (iUS) images. They organized a challenge for three tasks (pre-resection, during, and after). All medical images used for the challenge were acquired for routine clinical care at St Olavs University Hospital (Trondheim, Norway) after patients gave their informed consent. The imaging data are available in both MINC-2 and NIFTI-1 formats and the segmentations are available in NIFTI-1 format. All images, MRI, iUS, and segmentations are in the same referential space. A detailed description can be found [10, 11].

We have used the following preprocessing steps for data cleaning:

- Cropping strategy: Yes
- Resampling Method for anisotropic data: The nearest neighbor interpolation method has been applied for resampling.
- Intensity Normalization method:

The dataset has been normalized using a z-score method based on mean and standard deviation.

4 Implementation Details

The learning rate of 0.0004 with Adam optimizer has been for training the proposed model. The cross-entropy and dice function is used as a loss function between the output of the model and the ground-truth sample. 2 batch-size with 1000 epochs has been used with 20 early stopping steps. The best model weights have been saved for prediction in the validation phase. The 128 × 128 × 128 and others input image patches were used for training and the sliding window with stride 8 was used as the prediction. The Pytorch library is used for model development, training, optimization, and testing. The V100 tesla NVidia-GPU machine is used for training and testing the proposed model. The data augmentation methods mentioned in Table 1 are used for self-supervision stage 1 and the proposed model stage 2 for training and optimization. The dataset cases have different intensity ranges. The dataset is normalized between 0 and 1 using the max and min intensity normalization method. The detail of the training protocol is shown in Table 1.

Table 1. Training protocols.

Data augmentation methods	RandCrop, RandGaussianNoise, RandGaussianSmooth, RandShiftIntensity, RandAdjustContrast, RandZoomd
Initialization of the network	"he" normal initialization
Patch sampling strategy	None
Batch size	2
Patch size	128 × 128 × 128
Total epochs	300
Optimizer	Adam
Initial learning rate	0.0001
Learning rate decay schedule	None
Stopping criteria, and optimal model selection criteria	The stopping criterion is reaching the maximum number of epochs (300)
Training time	8 h

The same preprocessing has been applied at testing time. The training size of each image is fixed (128 × 128 × 128) and used linear interpolation method to resample the prediction mask to the original shape for each validation volume. The sliding window with has been used to get the prediction. The prediction mask produced by our proposed model has been resampled such that it has the same size and spacing as the original image and copies all of the meta-data, i.e., origin, direction, orientation, etc.

5 Results of a Validation Dataset

Figure 2 results in visualization of some validation cases. One bad and one good example for different subjects has been shown in Fig. 2. The proposed model achieved good performance in some cases and predicted a good segmentation map. The Dice value for before, during, and after tasks based on the proposed model is shown in Table 2. The dice score in a few cases is very bad, especially during and after the stage. We will try to optimize the model in the future to get a better Dice score. The results on test dataset is shown in Tables 3 and 4. We have evaluated our proposed model using different patch size. The proposed model with self-supervised module achieved better performance on task 1 and task 2.

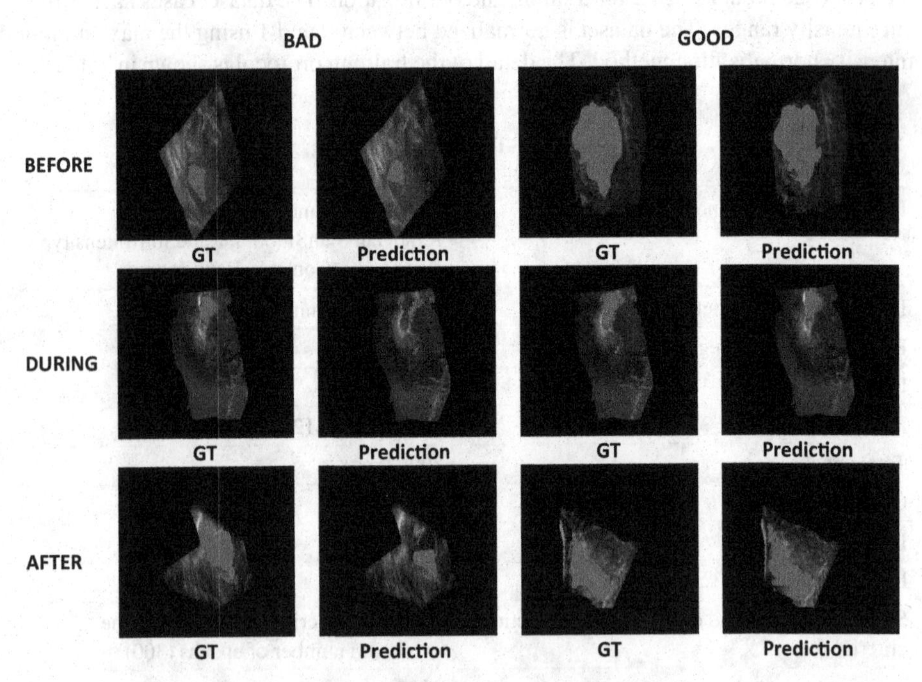

Fig. 2. 2D visualization using our proposed model.

Table 2. The performance of the proposed model on five validation cases for before, during, and after tasks

	S1	S2	S3	S4	S5	Average
Before	0.825905	0.898576	0.882049	0.310432	0.730627	0.729518
During	0.918594	0.769752	0.237226	0.505755	0.53811	0.593887
After	0.896442	0.895569	0.811527	0.055052	0.532311	0.63818

Table 3. The performance of proposed model using Task 1 (Brain tumor segmentation in intra-operative ultrasound) dataset

Models	Dice	Hd95	Recall	Precision
3DResUNet (128 × 128 × 128)	0.4769	99.69	0.5409	0.6230
3DResUNet + SSL (128 × 128 × 128)	0.5746	35.26	0.5845	0.7958
3DResUNet + SSL (64 × 64 × 64)	0.5430	42.40	0.5929	0.7829
3DResUNet + SSL (256 × 256 × 256)	0.5618	86.59	0.6275	0.5814

Table 4. Performance of proposed models using Task 2 (Resection cavity segmentation in intra-operative ultrasound) dataset

Models	Dice	Hd95	Recall	Precision
3DResUNet + SSL (128 × 128 × 128)	0.7027	24.54	0.6568	0.7829
3DResUNet + SSL (64 × 64 × 64)	0.6859	31.42	0.6213	0.8604
3DResUNet + SSL (256 × 256 × 256)	0.6791	22.66	0.6033	0.8686

6 Conclusion

In this work, we presented self-supervision segmentation of Intra-operative Ultrasound and presented 3D-ResUnet model with deep supervision. Experiments are conducted on s CuRIOUS-2022 showed significant better performance 0.729, 0.594 and 0.638 ((before, during, and after Intra-operative ultrasound (iUS) segmentation). The overall results on test dataset showed better performance and our proposed solution could be used as the first step towards correct diagnoses and prediction of iUS segmentation. In Future, we will develop 3D transformer-based model.

References

1. Payette, K., et al. : Fetal brain tissue annotation and segmentation challenge results. arXiv preprint arXiv:2204.09573 (2022)
2. Ma, J., et al.:Fast and Low-GPU-memory abdomen CT organ segmentation: The FLARE challenge. Medical Image Analysis (2022): 102616
3. Lalande, A., et al. "Deep Learning methods for automatic evaluation of delayed enhancement-MRI. The results of the EMIDEC challenge." Medical Image Analysis 79 (2022): 102428
4. Noreen, N., Palaniappan, S., Qayyum, A., Ahmad, I., Alassafi, M.O.: Brain tumor classification based on fine-tuned models and the ensemble method. Comput. Mater. Continua **67**(3), 3967–3982 (2021)
5. Chen, Z., et al.: Automatic deep learning-based myocardial infarction segmentation from delayed enhancement MRI. Comput. Med. Imaging Graph. **95**, 102014 (2022)
6. Chen, T., et al. "A simple framework for contrastive learning of visual representations." International conference on machine learning. PMLR, 2020

7. Pathak, D., et al. "Context encoders: Feature learning by inpainting." Proceedings of the IEEE conference on computer vision and pattern recognition. 2016
8. Chen, L., et al. "Self-supervised learning for medical image analysis using image context restoration." Medical image analysis 58 (2019): 101539
9. https://github.com/Project-MONAI/MONAI
10. Xiao, Y., Fortin, M., Unsgård, G., Rivaz, H., Reinertsen, I.: REtroSpective evaluation of cerebral Tumors (RESECT): a clinical database of pre-operative MRI and intra-operative ultrasound in low-grade glioma surgeries. Med. Phys. **44**(7), 3875–3882 (2017)
11. The dataset can be downloaded at the following address. https://archive.norstore.no/pages/public/datasetDetail.jsf?id=10.11582/2017.00004

Ultrasound Segmentation Using a 2D UNet with Bayesian Volumetric Support

Alistair Weld(✉), Arjun Agrawal(✉), and Stamatia Giannarou

The Hamlyn Centre, Imperial College London, London SW7 2AZ, UK
{a.weld20,arjun.a,stamatia.giannarou}@imperial.ac.uk

Abstract. We present a novel 2D segmentation neural network design for the segmentation of tumour tissue in intraoperative ultrasound (iUS). Due to issues with brain shift and tissue deformation, pre-operative imaging for tumour resection has limited reliability within the operating room (OR). iUS serves as a tool for improving tumour localisation and boundary delineation. Our proposed method takes inspiration from Bayesian networks. Rather than using a conventional 3D UNet, we develop a technique which samples from the volume around the query slice, and perform multiple segmentation's which provides volumetric support to improve the accuracy of the segmentation of the query slice. Our results show that our proposed architecture achieves an 0.04 increase in the validation dice score compared to the benchmark network.

Keywords: Ultrasound segmentation · 2D UNet · Bayesian

1 Introduction and Related Work

Intra-operative ultrasound (iUS) is a valuable tool for surgeons, providing access to real-time tissue characterisation. Pre-operative imaging such as magnetic resonance imaging (MRI) and computerised tomography (CT) are fundamental tools for the diagnosis of brain tumours. However, the occurrence of brain shift and tissue deformation [1] create discrepancies between what was previously captured and the current anatomical state. Adoption of these imaging tools into the operating room has been successful [2] but their cumbersome nature renders them infeasible for real-time tissue analysis. iUS on the other hand is lightweight and requires minimal preparation. Due to the readiness of data, the majority of research has been conducted on MRI and CT images. The main datasets for these modalities are BRATS [3], MSD [4] and KiTS [5]; [6] provides a complete overview of the application of deep learning in medical imaging (predominantly MRI). In this work, we are solving the task of iUS tumour segmentation using a novel architecture, which is designed to achieve good performance on small training datasets.

Y. Xiao et al. (Eds.): CuRIOUS/KiPA/MELA 2022, LNCS 13648, pp. 63–68, 2023.
https://doi.org/10.1007/978-3-031-27324-7_8

2 Method

2.1 Network Architecture

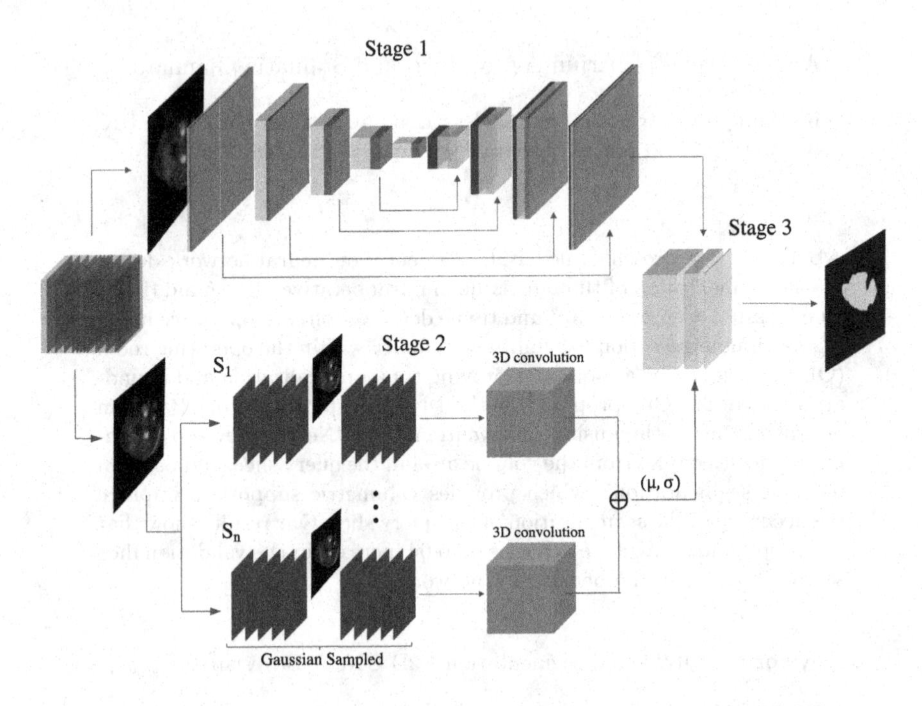

Fig. 1. Proposed model architecture used to segment tumour cavities in iUS scans.

Since its inception, the UNet architecture [7] has been the most commonly used neural network foundation for medical image segmentation; relevant examples of this [8–10]. A model has been designed with three components - (1) A standard UNet is used to provide a provisional estimation of the 2D segmentation mask of the query image defined by f_a, using only the 2D slice as input (2) A sequence of 4 3D convolution blocks is used to provide multiple 3D segmentations of the volume around the query image defined by f_b - 2 sub-volumes are used in our work, however, this can be set to any number and can also be changed during validation (3) The outputs of (1) and (2) are concatenated and processed through a sequence of 4 2D convolutions to provide the final segmentation mask defined by f_c. These stages are mathematically defined in Eq. 2 and displayed in Fig. 1. The query image, $v^{query} \in V \in \mathbb{N}^{320,320}$, is the specific ultrasound slice which is being segmented. A Gaussian distribution is used to sample 5 separate sets of sub-volumes with 10 samples each, where the mean (μ) is the query slice's index and the distribution's standard deviation (σ) is 8 as shown in Eq. 1. v^{query} is then processed through stage 1. Each sub-volume, $v_s^{supp} \in \mathbb{N}^{320,320}$, is individually segmented using the same 3D convolution sequence - stage 2. Skip connections containing the original input volume are concatenated in between each convolutional layer. The query images are then extracted from each subvolume and the mean and standard deviation are calculated per pixel. This process means that inferencing can be achieved on more subvolumes than was used during training. The outputs of stages 1 and 2 are then concatenated and processed through stage 3, which produces the final output. Skip connections containing v^{query} are concatenated after each convolutional layer. All inputs to the neural network are down-sampled to $\mathbb{N}^{320,320}$ using nearest neighbour resampling - same dimension for training and optimisation. For the evaluation, $\hat{P}$ is upsampled using bilinear resampling.

The design of stage 2 takes inspiration from [11]. However, instead of applying the dropout on the neural network, dropout is applied on the 3D volume. Where the v^{query} is the central slice in each subvolume. This provides different volumetric perspectives around the v^{query}. The Gaussian sampling also provides a regularizing effect on the training. This is an alternative design to simply using a 3D UNet, which more efficiently uses the training data.

The network is optimised using a weighted combination loss consisting of binary cross entropy and the Dice coefficient. Each stage in the proposed architecture is designed to produce a segmentation and is optimised using binary cross entropy and the Dice coefficient - as show in Eq. (3)–(5). Stage 1 provides the initial estimation of the mask. Stage 2 provides an estimation of the 3D region around the mask. Stage 3 generates the final segmentation prediction by combining the information provided by the 2D segment and the mean and standard deviation of the volumetric prediction.

$$v_s^{supp} = x_1, ..., x_n \overset{\text{i.i.d}}{\sim} \mathcal{N}(\mu, \sigma^2), s \in [1, 2, 3, 4, 5] \tag{1}$$

$$\hat{P} = f_c(f_a(v^{query}) \oplus \bigodot_{s=0}^{0} f_b(v_s^{supp})) \tag{2}$$

$$BCE = -(y \log(\hat{y}) + (1 - y) \log(1 - \hat{y})) \tag{3}$$

$$Dice = \frac{2 \sum_i y_i \hat{y}_i + 1}{\sum_i y_i + \sum_i \hat{y}_i + 1} \tag{4}$$

$$\begin{aligned} L = BCE(\hat{P}) + \lambda_a * Dice(\hat{P}) + \lambda_b * (Dice(f_b(v_s^{supp}) \\ +BCE(f_b(v_s^{supp}) + Dice(f_a(v^{query}) + BCE(f_a(v^{query})))) \end{aligned} \tag{5}$$

3 Experiments

3.1 Training Setup and Strategy

Our model was benchmarked against the standard UNet [7] - a submodule of our proposed architecture. All training was done using only the provided RESECT [12] dataset. The neural network was trained using a Linux machine containing an NVIDIA RTX A5000 GPU and an 11th Generation Intel(R) Core(TM)

Table 1. Performance of tested models across the 4 evaluation metricises

Cases	Training			
	Dice	Hausdorf	Accuracy	Recall
2D UNet	0.607	42.603	0.916	0.588
Proposed	0.773	26.393	0.846	0.625
	Validation			
	Dice	Hausdorff	Accuracy	Recall
2D UNet	0.639	35.264	0.857	0.600
Proposed	0.677	41.606	0.700	0.612

i9-11900K @ 3.50 GHz. The PyTorch [13] machine learning framework was used to support the training. Both models were trained for 10 epochs with a batch size of 8.

Due to its general robustness, the ADAM [14] optimizer was used. The initial learning rate is set to 10^{-5}. In order to achieve better generalisability, the number of training samples were increased by utilising a set of data augmentation techniques - thus widening the training distribution. The TorchIO [15] library was used as it has been designed for medical image augmentation. The used augmentations were: RandomAffine, RandomMotion, RandomGhosting, RandomSpike, RandomBiasField, RandomBlur, RandomNoise.

3.2 Results

The dataset was split into training/validation using the ratio 85:15, where 18 cases were used for training and 3 cases (25, 26 and 27) were used for validation. In order to show the benefit of the proposed method, we compare our performance with the standard 2D UNet. Our results are shown in Table 1 - it is observed that our proposed method improves the Dice score and recall whilst the baseline produces better results for the Hausdorff score and accuracy. We note that the accuracy of a predicted segmentation is not an ideal evaluation metric for medical segmentation tasks, as these tasks often suffer from a large class imbalence - a bias caused by the larger number of background labels than tumour e.g. Fig. 2. Hence, it is possible for a prediction to have high accuracy and poor tumour segmentation.

Shown in Fig. 2 is an example of predictions made by the baseline and proposed method, compared to the ground truth. Though a small increase in the dice score, it is important to highlight the increased boundary prediction accuracy in the proposed method. The delineation of healthy tissue and pathological tissue is possibly the most important cause for application of iUS.

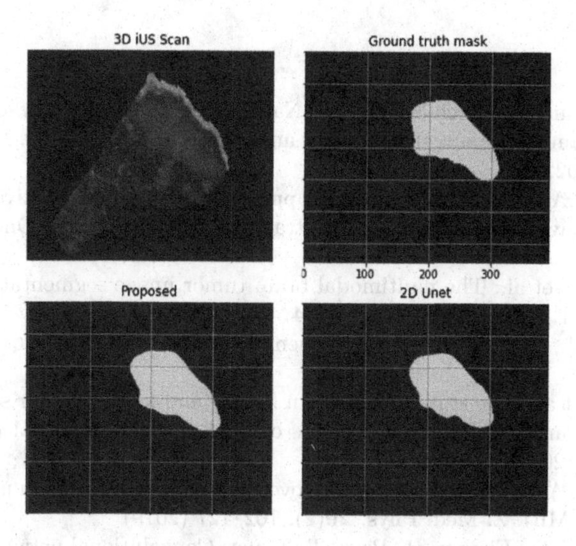

Fig. 2. Comparison of proposed model architecture and baseline 2D UNet, used to segment tumour cavities in iUS scans.

4 Conclusion

Our work presents a novel approach to segmenting 3D volumetric medical scans with small datasets. We develop on the established 2D UNet by adding a volumetric 3D sub-network, which is used to support the 2D prediction. Our evaluation shows that this addition improves the performance, compared to the baseline 2D UNet, while not exponentially increasing compute. Due to the small nature of the RESECT dataset, using a 3d UNet would limit the number of optimizing steps. Hence, with the volumetric support system, our solution is able to leverage the efficiency of 3D convolutions without limiting training data. An interesting further research direction could be the exploration into how best to sample from the 3D volume and combine that information with the 2D UNet. For simplicity, we limited our work to only sampling from the close neighbouring area around the query image. However, the sampling could involve multiple σ values. We also used a simple sequence of 3D convolutions to process the volumetric data, this could be improved on. We also hypothesise that our performance could have been improved if we had developed a per image classification network, designed to classify if an image contains a tumour or not. This could help to prevent false positive and false negative predictions.

Acknowledgements. This project was supported by UK Research and Innovation (UKRI) Centre for Doctoral Training in AI for Healthcare (EP/S023283/1).

References

1. Dixon, L., Lim, A., Grech-Sollars, M., Nandi, D., Camp, S.J.: Intraoperative ultrasound in brain tumor surgery: a review and implementation guide. Neurosurg. Rev. **45**, 1–13 (2022)
2. Bastos, D.C.A., et al.: Challenges and opportunities of intraoperative 3d ultrasound with neuronavigation in relation to intraoperative MRI. Front. Oncol. **11**, 656519 (2021)
3. Menze, B.H., et al.: The multimodal brain tumor image segmentation benchmark (brats). IEEE Trans. Med. Imaging **34**, 1993–2024 (2015)
4. Antonelli, M., et al.: The medical segmentation decathlon. Nat. Commun. **13**, 4128 (2022)
5. Heller, N., et al.: The state of the art in kidney and kidney tumor segmentation in contrast-enhanced CT imaging: results of the kits19 challenge. Med. Image Anal. **67**, 101821 (2020)
6. Lundervold, A.S., Lundervold, A.: An overview of deep learning in medical imaging focusing on MRI. Z. Med. Phys. **29**(2), 102–127 (2019)
7. Ronneberger, O., Fischer, P., Brox, T.: U-net: Convolutional networks for biomedical image segmentation. arXiv:1505.04597 (2015)
8. Canalini, L., Klein, J., Miller, D., Kikinis, R.: Enhanced registration of ultrasound volumes by segmentation of resection cavity in neurosurgical procedures. Int. J. Comput. Assist. Radiol. Surg. **15**, 1963–1974 (2020)
9. Canalini, L., et al.: Segmentation-based registration of ultrasound volumes for glioma resection in image-guided neurosurgery. Int. J. Comput. Assist. Radiol. Surg. **14**, 1697–1713 (2019)
10. Zhong, X., et al.: Deep action learning enables robust 3d segmentation of body organs in various CT and MRI images. Sci. Rep. **11**, 3311 (2021)
11. Gal, Y., Ghahramani, Z.: Dropout as a Bayesian approximation: representing model uncertainty in deep learning. arXiv:1506.02142 (2016)
12. Xiao, Y., Fortin, M., Unsgård, G., Rivaz, H., Reinertsen, I.: Retrospective evaluation of cerebral tumors (resect): a clinical database of pre-operative MRI and intra-operative ultrasound in low-grade glioma surgeries. Med. Phys. **44**(7), 3875–3882 (2017)
13. Paszke, A., et al.: Pytorch: an imperative style, high-performance deep learning library. In: Advances in Neural Information Processing Systems, vol. 32, pp. 8024–8035. Curran Associates Inc. (2019). https://papers.neurips.cc/paper/9015-pytorch-an-imperative-style-high-performance-deep-learning-library.pdf
14. Kingma, D.P., Ba, J.: Adam: A method for stochastic optimization. CoRR abs/1412.6980 (2015)
15. Pérez-García, F., Sparks, R., Ourselin, S.: TorchIO: a python library for efficient loading, preprocessing, augmentation and patch-based sampling of medical images in deep learning. Comput. Methods Progr. Biomed. **208**, 106236 (2021)

Segmentation of Intraoperative 3D Ultrasound Images Using a Pyramidal Blur-Pooled 2D U-Net

Mostafa Sharifzadeh[1,2(✉)], Habib Benali[1,2], and Hassan Rivaz[1,2]

[1] Department of Electrical and Computer Engineering, Concordia University, Montreal, QC H3G 1M8, Canada
{hassan.rivaz,habib.benali}@concordia.ca
[2] PERFORM Centre, Concordia University, Montreal, QC H4B 1R6, Canada
mostafa.sharifzadeh@concordia.ca

Abstract. Automatic localization and segmentation of the tumor and resection cavity in intraoperative ultrasound images can assist in accurate navigation during image-guided surgery. In this study, we benchmark a pyramidal blur-pooled 2D U-Net as a baseline method to segment the tumor and resection cavity before, during, and after resection in 3D intraoperative ultrasound images. Slicing the 3D image along three transverse, sagittal, and coronal axes, we train a different model corresponding to each axis and average three predicted masks to obtain the final prediction. It is demonstrated that the averaged mask consistently achieves a Dice score greater than or equal to each individual mask predicted by only one model along one axis.

Keywords: Neural networks · Segmentation · Ultrasound · U-net

1 Introduction

The success rate of safely resecting a brain tumor during neurosurgery highly depends on an accurate and reliable intraoperative neuronavigation [1]. Preoperative imaging methods such as magnetic resonance (MR) imaging play a pivotal role in neurosurgery; however, distortions, deformations, and brain shifts make those images less valuable during the operation. Intraoperative ultrasound is an affordable, safe, and real-time imaging technique that, due to its high temporal resolution, can be easily incorporated into the surgical workflow and provides live imaging during surgery. Although ultrasound images are more difficult to be interpreted compared to other modalities such as MR images, automatic segmentation of intraoperative ultrasound images provides an effective tool to mitigate the issue by, for instance, facilitating registration of preoperative MR and intraoperative ultrasound images. In this work, we employed a pyramidal blur-pooled 2D U-Net to perform two tasks requested by the CuRIOUS 2022 - Segmentation Challenge organizers: segmentation of brain tumor in pre-resection 3D ultrasound images (Task 1) and segmentation of resection cavity in post-resection 3D ultrasound images (Task 2).

Y. Xiao et al. (Eds.): CuRIOUS/KiPA/MELA 2022, LNCS 13648, pp. 69–75, 2023.
https://doi.org/10.1007/978-3-031-27324-7_9

2 Methodology

2.1 Dataset

We used RESECT, the publicly available dataset, including preoperative contrast-enhanced T1-weighted and T2 FLAIR MRI scans alongside three 3D volumes of intraoperative ultrasound scans acquired from 23 clinical patients with low-grade gliomas before, during, and after undergoing tumor resection surgeries [2]. In this study, only the ultrasound volumes were employed for the segmentation tasks, where delineations of the brain tumors and resection cavities had been provided as ground truths in addition to the original database [3].

We split the dataset into training and validation sets containing 19 and 4 cases, respectively. Seven additional cases, provided by the challenge organizers, were used as the test set. All volumes were normalized between 0 and 1, individually, then zero-padded symmetrically to the maximum size existing in the dataset along each axis and finally resampled to the size of 150×150×150.

2.2 Network Architecture

The U-Net [4] was proposed for biomedical segmentation applications wherein the training sets are typically small. It consists of an encoder followed by a decoder, in which the former extracts locality features and the latter resamples the image maps with contextual information. To produce more semantically meaningful outputs, it includes skip connections at each layer for concatenating low- and high-level features. Since this architecture has shown promising results and is the most commonly used one in this domain, we also employed a pyramidal blur-pooled U-Net [5], a variant of this network that is more robust to the shift-variance problem [6] and provides higher output consistency. Compared to the vanilla U-Net, the max-pooling layers are replaced with blur-pooling layers in pyramidal blur-pooled U-Net. The anti-aliasing filters in pyramidal blur-pooled were of sizes 7×7, 5×5, 3×3, and 2×2, from the first to the forth downsampling layer, respectively.

Let $\boldsymbol{I} \in \mathbb{R}^d$ and $\hat{\boldsymbol{S}} \in \{0,1\}^d$ denote a sample input image and the corresponding predicted mask, respectively. $f_{seg} : \mathbb{R}^d \rightarrow \{0,1\}^d$ is the network that predicts the segmentation mask by accepting the input image.

$$\hat{\boldsymbol{S}} = f_{seg}(\boldsymbol{I}, \boldsymbol{\theta}) \tag{1}$$

where $\boldsymbol{\theta}$ are the network's parameters.

2.3 Training Strategy

For each task, we trained three different models using the 2D images acquired by slicing 3D volumes along three transverse, sagittal, and coronal axes. By slicing 3D volumes, we obtained a highly imbalanced dataset wherein many images had a completely black mask (no foreground). To mitigate this issue and to achieve a faster training time, we trained the models merely using images with a non-zero mask (including at least one pixel as the foreground) and discarded the rest. However, even in this case, the dataset was still imbalanced as a large majority of the pixels were background in the remaining masks. To alleviate this problem, we employed the focal Tversky loss function which is a generalized focal loss function based on the Tversky index and was proposed to address the issue of data imbalance in medical image segmentation [7].

For Task 1 experiments, we merely used before-resection volumes; however, for Task 2 experiments, both during, and after-resection volumes were combined and considered as one dataset. Task 2 experiments were trained from scratch; whereas Task 1 experiments were initialized using pre-trained weights obtained from Task 2.

The sigmoid function was employed as the activation function of the last layer, and the batch size was 32. We utilized AdamW [8] as the optimizer, and set the weight decay parameter to 10^{-2}. We trained each network for 500 epochs, and saved model weights only if the validation loss had been improved and finally used the best weights for testing. The learning rate was set to 2×10^{-4} initially and was lowered by 2 times at epochs 300 and 400. The same configuration was used for all experiments. They were implemented using the PyTorch package [9], and training was performed on two NVIDIA A100 GPUs utilizing the DataParallel class, which parallelizes the training by splitting the input across the two GPUs by chunking in the batch dimension.

2.4 Augmentation

During the training, six on-the-fly augmentation techniques were applied. We randomly scaled the 2D images by $s\%$ along both axes, where $s \in [-7, +7]$. We also applied a Gaussian smoothing filter with a kernel of size $k \times k$ and standard deviation σ, where $k \in \{0, 2, 3, 5\}$ and $\sigma \in [0, 0.6]$ were chosen randomly. Besides, we randomly altered images' brightness and contrast and flipped (horizontally) and rotated (θ *degrees*) them, where the chance of flipping was 50%, and $\theta \in [-15, 15]$. As the dominant noise source in ultrasound images, we modeled the speckle noise as a multiplicative noise and randomly applied it to the images:

$$\boldsymbol{I}_{noisy} = \boldsymbol{I} + \boldsymbol{N}\boldsymbol{I} \tag{2}$$

where $\boldsymbol{N}$ is a matrix with the same size of the image consisting of normally distributed values with zero mean and standard deviation $\sigma \in [0, 0.01]$.

Finally, we randomly cropped a patch of size 128×128 from images of the original size 150×150, which is equivalent to the translation augmentation. Since choosing and storing the best model was based on the validation set, we always used center-copped images without any augmentations during the validation phase.

3 Results

To predict the segmentation mask of each 3D volume, we followed the preprocessing procedure same as for the training. The volume was normalized between 0 and 1, zero-padded symmetrically to the maximum size existing in the dataset along each axis, resampled to the size of 150×150×150, and center-cropped to 128×128×128. Then it was sliced along the three axes, and 2D slices along each axis were fed into the corresponding network to predict 2D masks. Afterward, 2D masks were stacked together to form three 3D volumes, each corresponding to one of the axes. Finally, we averaged three volumes and thresholded the resulting volume at 0.5. To make sure that the mask size matches the original image size, it was zero-padded symmetrically to the size of 150×150×150 and resampled to the size of the original image volume to obtain the final predicted mask.

Applying the method to before, during, and after resection volumes of all cases in the validation set, Table 1 shows the resulted Dice scores according to the predicted masks and ground truths. Although the dataset contained 23 cases, note that the cases in the table are labeled based on the original labels in the publicly available dataset, wherein they were not necessarily numbered consecutively. According to Table 1, an easy (#26) and a difficult (#24) case are chosen, and the qualitative results are demonstrated in Fig. 1. It shows the results for Task 1, where the tumor is segmented in a pre-resection image, and Task 2, where the resection cavity is segmented in a post-resection image. Finally, the performance across the test set is summarized in Table 2 based on the averaged Dice, HD95, recall, and precision metrics provided by the challenge organizer after submitting the results.

Table 1. Dice scores of the validation set, where B, D, and A stand for before, during, and after resection, respectively, with the highest values shown in bold.

	Case #24			Case #25			Case #26			Case #27		
Stage	B	D	A	B	D	A	B	D	A	B	D	A
Axis 0	0.25	0.13	0.87	0.66	0.83	0.88	0.87	0.89	0.95	0.70	0.11	0.93
Axis 1	0.27	0.1	0.83	0.63	0.78	0.84	0.91	0.85	0.96	0.76	0.16	0.94
Axis 2	0.27	0.05	0.86	0.64	0.78	0.87	0.90	0.88	0.90	0.78	0.11	0.84
Final	**0.47**	**0.23**	**0.88**	**0.73**	**0.89**	**0.92**	**0.92**	**0.93**	**0.96**	**0.85**	**0.21**	**0.94**

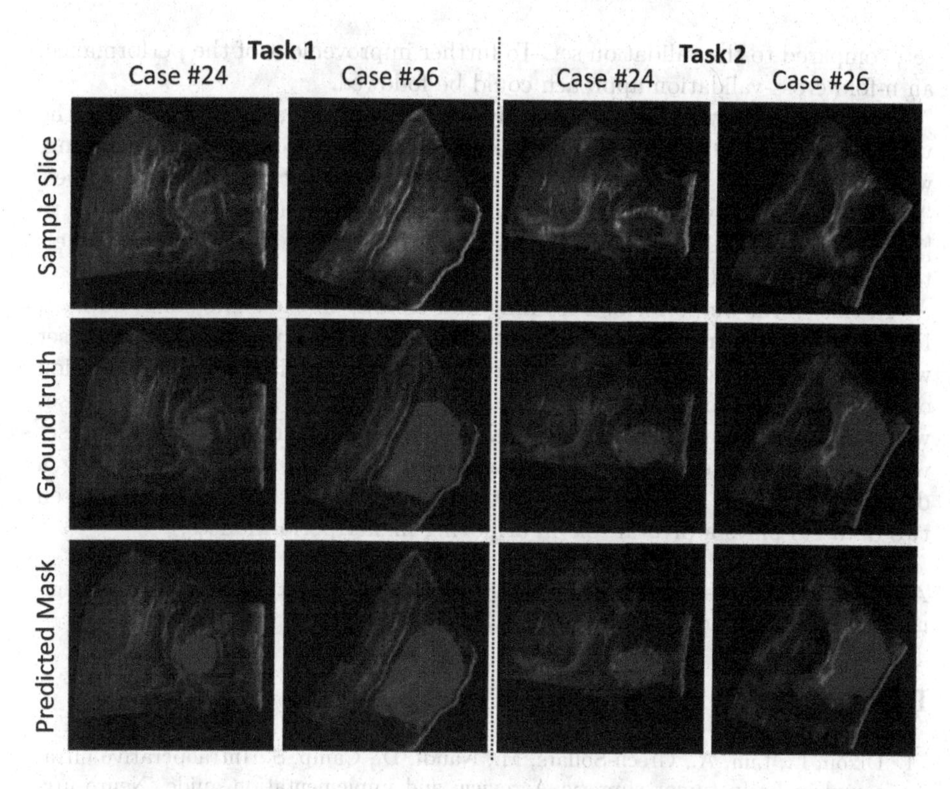

Fig. 1. Sample slices from two cases of the validation set. Cases #24 and #26 represent the lowest and highest dice scores, respectively. The first and second columns show samples of Task 1, where the tumor is segmented in a pre-resection image, and the third and fourth columns correspond to Task 2, where the resection cavity is segmented in a post-resection image.

Table 2. Performance of the method across the test set.

	Dice	HD95	Recall	Precision
Task 1	0.53	71.57	0.64	0.54
Task 2	0.62	36.08	0.54	0.80

4 Discussion and Conclusions

We benchmarked a pyramidal blur-pooled 2D U-Net as a baseline method to segment the tumor and resection cavity in 3D intraoperative ultrasound images. To this end, we predicted three different masks for each volume based on the slicing axis and averaged them before thresholding to obtain the final mask. In Table 1, we can see that the final mask consistently achieved a Dice score greater than or equal to each individual mask predicted by only one model along one axis. In Table 2, a drop in performance can be observed for the test set, compared to the validation set. To further improvement of the performance, an n-fold cross-validation approach could be followed.

One of the drawbacks of this study was that we separated 4 cases as the validation set and those cases were never taken into account for the training, which means losing a large portion of data in a small dataset of only 23 cases. Another limitation was utilizing only one slice as the input; however, similar to [10] [11], adjacent slices also could be fed into the network as the input channels to improve the predictions.

In this work, we performed training and testing on the same small dataset. However, higher performance is expected by augmenting the small training set with more annotated data. For instance, an ultra-fast method for the simulation of realistic ultrasound images [12] could be employed to augment the dataset with simulated data. Another approach would be augmenting the training set with other annotated ultrasound datasets. In these approaches, employing a domain adaptation method [13] to mitigate the domain shift problem between two datasets plays a pivotal role in achieving higher performances.

Acknowledgements. We acknowledge the support of the Natural Sciences and Engineering Research Council of Canada (NSERC).

References

1. Dixon, L., Lim, A., Grech-Sollars, M., Nandi, D., Camp, S.: Intraoperative ultrasound in brain tumor surgery: A review and implementation guide. Neurosurg. Rev. **45**(4), 2503–2515 (2022)
2. Xiao, Y., Fortin, M., Unsgärd, G., Rivaz, H., Reinertsen, I.: REtroSpective Evaluation of Cerebral Tumors (RESECT): A clinical database of pre-operative MRI and intra-operative ultrasound in low-grade glioma surgeries: A. Med. Phys. **44**(7), 3875–3882 (2017)
3. Behboodi, B., et al.: RESECT-SEG: Open access annotations of intra-operative brain tumor ultrasound images. (2022)
4. Ronneberger, O., Fischer, P., Brox, T.: U-net: Convolutional networks for biomedical image segmentation. Lect. Notes Comput. Sci. **9351**, 234–241 (2015)
5. Sharifzadeh, M., Benali, H., Rivaz, H.: Investigating Shift Variance of Convolutional Neural Networks in Ultrasound Image Segmentation. IEEE Trans. Ultrason. Ferroelectr. Freq. Control **69**(5), 1703–1713 (2022)
6. Sharifzadeh, M., Benali, H., Rivaz, H.: Shift-Invariant Segmentation in Breast Ultrasound Images. IEEE International Ultrasonics Symposium, IUS (2021)
7. Abraham, N., Khan, N.M.: A novel focal Tversky loss function with improved attention u-net for lesion segmentation. In: Proceedings - International Symposium on Biomedical Imaging, (ISBI), pp. 683–687 (2019)
8. Loshchilov, I., Hutter, F.: Decoupled weight decay regularization. In: 7th International Conference on Learning Representations, ICLR (2019)
9. Adam Paszke, A.: PyTorch: an imperative style, high-performance deep learning library. In Adv. Neural Inf. Proc. Syst. **32** (2019)
10. Carton, F.-X., Chabanas, M., Le Lann, F., Noble, J.H.: Automatic segmentation of brain tumor resections in intraoperative ultrasound images using U-Net. J. Med. Imaging **7**(03), 1 (2020)

11. Carton, F.-X., Noble, J.H., Chabanas, M.: Automatic segmentation of brain tumor resections in intraoperative ultrasound images. In Fei, B., Linte, C.A., eds, Medical Imaging 2019: Image-Guided Procedures, Robotic Interventions, and Modeling. Society of Photo-Optical Instrumentation Engineers(SPIE) 7, p. 104 (2019)
12. Sharifzadeh, M., Benali, H., Rivaz, H.: An Ultra-Fast Method for Simulation of Realistic Ultrasound Images. IEEE International Ultrasonics Symposium, IUS (2021)
13. Sharifzadeh, M., Tehrani, A.K.Z., Benali, H., Rivaz, H.: Ultrasound Domain Adaptation Using Frequency Domain Analysis. In: IEEE International Ultrasonics Symposium (IUS), pp. 1–4(2021)

The 2022 Mediastinal Lesion Analysis Challenge (MELA 2022)

Accurate Detection of Mediastinal Lesions with nnDetection

Michael Baumgartner[1(✉)], Peter M. Full[1,2], and Klaus H. Maier-Hein[1,3]

[1] Division of Medical Image Computing, German Cancer Research Center, Heidelberg, Germany
m.baumgartner@dkfz.de

[2] Medical Faculty Heidelberg, Heidelberg University, Heidelberg, Germany

[3] Pattern Analysis and Learning Group, Heidelberg University Hospital, Heidelberg, Germany

Abstract. The accurate detection of mediastinal lesions is one of the rarely explored medical object detection problems. In this work, we applied a modified version of the self-configuring method nnDetection to the Mediastinal Lesion Analysis (MELA) Challenge 2022. By incorporating automatically generated pseudo masks, training high capacity models with large patch sizes in a multi GPU setup and an adapted augmentation scheme to reduce localization errors caused by rotations, our method achieved an excellent FROC score of 0.9922 at IoU 0.10 and 0.9880 at IoU 0.3 in our cross-validation experiments. The submitted ensemble ranked third in the competition with a FROC score of 0.9897 on the MELA challenge leaderboard.

Keywords: Mediastinal lesions · CT · Object detection · nnDetection

1 Introduction

While many pathologies were already explored for medical object detection by previous studies [1,4,6,8], the detection of lesions in the mediastinum was rarely investigated before. The MELA challenge 2022 aims to address this shortcoming by providing a large, publicly available data set and leaderboard to benchmark new advances in the domain. Our submission to the challenge is based on nnDetection [1], a recently proposed self-configuring object detection method which can be applied to new problems without manual intervention. It follows nnU-Net's [3] design principle to systematise and automate the hyperparameter tuning process by using fixed, rule-based and empirically derived parameters. Since nnDetection was developed for data sets with pixel-wise annotations and GPUs with 11GB of VRAM, we make minor adjustments to its configuration to fully leverage our available resources and tackle three challenges of this specific data set:

– **Bounding Box Annotations:** Enclosing lesions with bounding boxes, enables the collection of large scale data sets by speeding up the annotation process. However, compared to pixel-wise segmentation, less information

Y. Xiao et al. (Eds.): CuRIOUS/KiPA/MELA 2022, LNCS 13648, pp. 79–85, 2023.
https://doi.org/10.1007/978-3-031-27324-7_10

is stored in bounding boxes which limits the use of spatial data augmentation transformations and available training signals.
- **Large Lesions:** Lesion bounding boxes can only be predicted accurately when the entire lesion is visible in the image patches used for inference. MELA has particularly large lesions, requiring large patch sizes and in turn an immense amount of GPU memory to train appropriate models.
- **Accurate Localization:** MELA has an unusually high IoU cutoff at 0.3 (versus the commonly used 0.1 [1,5]) requiring particularly precise bounding box localization especially in the case of small lesions.

Our proposed solution incorporates automatically generated pseudo masks which were derived from the provided bounding boxes, training high capacity models with large patch sizes in a multi GPU setup and an adapted augmentation scheme to reduce localization errors caused by rotation when training with coarse annotations to tackle the aforementioned challenges.

2 Methods

We used nnDetection [1], a self-configuring method for volumetric medical object detection, as our development framework. Some of the fixed parameters were adapted to account for the availability of improved resources and the bounding box annotations while the rule based and empirical parameters were used without modification, see Sect. 2.1. Since many lesions exceed the patch size of the full resolution models, all of our experiments used the automatically created low resolution data set [1].

Preprocessing. Motivated by previous studies by Tang et al. [9] and Zlocha et al. [11] additional segmentation masks were generated to encode prior information into the training process. We used the center and size of the bounding boxes to fit ellipsoids which mimic the round shape of lesions. By leveraging the generated segmentations throughout the whole data preparation and data loading pipeline, we can use spatial transformations from commonly available data augmentation frameworks, such as batchgenerators [2], to artificially diversify the data. Since multiple CT scans in the data set had a similar spacing, the resampling step was only executed on scans where at least one of the spacings differed more than 5% from the target spacing of $[1.40, 1.43, 1.43]mm$.

Training. Each network is trained for 50 epochs each consisting of 2500 batches. Gradient updates are performed with Stochastic Gradient Descent (SGD) and a Nesterov momentum term of 0.99. During the data loading process, a random offset is applied to each dimension where the object size does not exceed 70% of the patch size while making sure that the object remains fully contained inside the patch. The maximum offset is limited to 70% of the difference between the object size and patch size. For object dimensions which exceed that threshold, a random center point is selected inside the object boundaries.

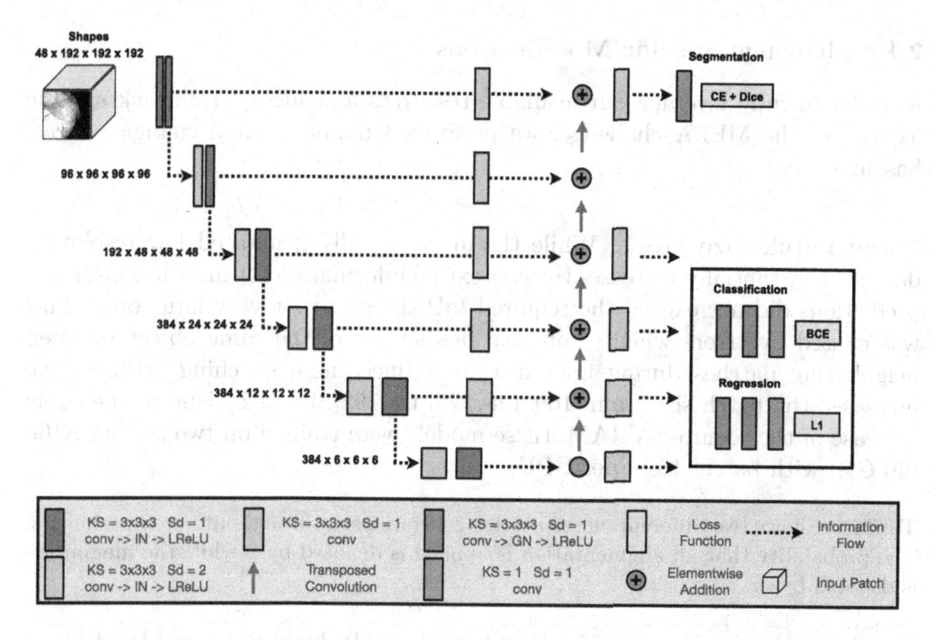

Fig. 1. Depicts Retina U-Net, the blueprint architecture of nnDetection which is automatically configured based on the data set properties. The patch size was manually adjusted to account for the large objects present in the MELA data set. The encoder architecture is based on stacked convolutions while the decoder is based on the Feature Pyramid Network [7]. Instead of bi-linear interpolation, transposed convolutions are used to learn the up-sampling. Abbreviations: conv = Convolution, KS = Kernel Size, Sd = Stride, IN = Instance Normalisation, GN = Group Normalisation [10], LReLU = Leaky Rectified Linear Unit

Network Topology. The network topology follows the original nnDetection implementation and consists of convolutions with non-linear activation and normalisation layers. The ReLU activations were replaced with Leaky ReLU activations to avoid dead neurons during the training process. We kept Retina U-Net [5] as our blueprint architecture for all experiments which results in three training branches: anchor classification trained with the Binary Cross Entropy loss, anchor regression trained with a weighted L1 loss and a semantic segmentation branch which is trained with the Cross Entropy and Dice loss. The total loss composition can be summarised as follows:

$$L_{total} = L_{BCE} + 2 \cdot L_{L1} + L_{CE\_seg} + L_{Dice\_seg} \quad (1)$$

To increase the model capacity, the number of channels were increased by 50% and the maximum number of channels was set to 384. A detailed overview of the resulting architecture is shown in Fig. 1.

2.1 Challenge Specific Modifications

In order to fully leverage our available resources and incorporate task specific aspects of the MELA challenge, we performed minor manual changes to our baseline.

Large Patch Size (LP). While the automatically generated low resolution data set substantially increases the contextual information for most lesions, some predictions did not exceed the required IoU threshold for very large ones. This was caused by errors when combining predictions of the same object between neighboring patches during inference. To reduce these stitching artifacts, we increased the patch size from [160, 128, 128] to [192, 192, 192]. Due to the cubic increase in the required VRAM, these models were trained on two Nvidia A100 (40 GB) with batch size 4 per GPU.

Table 1. Shows two different augmentation schemes used throughout our experiments. The probability that an augmentation is applied is denoted by p while the magnitude is denoted by m.

Augmentation	Baseline (Aug A)	Reduced Rotation (Aug B)
Elastic Deformation	×	×
Rotation (m in degrees)	p = 0.3 m = [−30, 30]	p = 0.1 m = [−10, 10]
Scaling	p = 0.2 m = [0.7, 1.4]	p = 0.3 m = [0.65, 1.6]
Rotation 90	×	p = 0.5
Transpose Axes	×	p = 0.5
Gaussian Noise	p = 0.1	p = 0.1
Gaussian Blur	p = 0.2	p = 0.2
Median Filter	×	p = 0.2
Multiplicative Brightness	p = 0.15	×
Brightness Gradient	×	p = 0.3
Contrast	p = 0.15	p = 0.2
Siumlate Low Resolution	×	p = 0.15
Gamma	p = 0.3	p = 0.1
Inverse Gamma	p = 0.1	p = 0.1
Local Gamma	×	p = 0.3
Sharpening	×	p = 0.2
Mirror (per axes)	p = 0.5	p = 0.5

Reduced Rotation During Augmentation. Spatial data augmentations are a central part of modern deep learning pipelines to reduce overfitting and improve generalisation of the developed methods. By applying rotations and scaling to the images, the diversity in the data set is artificially increased by training on augmented versions of the same patch. Rotations which are not a multiple of 90°C can be troublesome when coarse annotations are used during training since the axis aligned bounding boxes can not be derived without introducing localization errors. To account for this fact, a second augmentation pipeline with reduced rotation transformations was used to train a second model. By adding additional intensity based augmentations, as well as transposing and rotation of 90°C we tried to maintain a diverse set of augmentations. The full list of the augmentations, probabilities and magnitudes is shown in Table 1.

3 Experiments and Results

The MELA data set provides 770 training and 110 validation CT scans. To reduce overfitting during the development phase, all scans were merged into one pool and 5-fold cross-validation was used to train and evaluate the models. Performance is measured by the FROC score with sensitivities computed at $0.125, 0.25, 0.5, 1, 2, 4, 8$ False Positives per Image at an Intersection over Union (IoU) threshold of 0.3. Additionally, we report results at an IoU threshold of 0.1 for our cross-validation experiments. During the final phase, the test data was released to the participants and up to five submission per day were allowed. Our final submission was based on the ensemble of the two training strategies with large patch sizes and different data augmentation schemes. An overview of all the results can be found in Table 2.

Table 2. Results of different training strategies for our 5-fold cross-validation experiments and leaderboard submissions. Differences in patch size, effective batch size and data augmentation are listed in separate columns. The LP ensemble denotes the ensemble of the two training strategies with large patch sizes.

Model	Patch Size	Batch Size	Augmentation	5 Fold CV		Leaderboard
				IoU 0.1	IoU 0.3	
Baseline	[160,128,128]	6	A	0.9844	0.9808	0.9824
LP	[192,192,192]	8	A	0.9858	0.9809	0.9851
Aug B, LP	[192,192,192]	8	B	**0.9927**	0.9846	0.9851
LP Ensemble	[192,192,192]	8	A+B	0.9922	**0.9880**	**0.9897**

All models achieved an excellent FROC score above 0.98 while a performance gap between the lower and higher IoU thresholds persisted throughout our cross-validation experiments. Using a large patch size for additional contextual information showed slight improvements during the cross-validation and the leaderboard. Introducing the second augmentation scheme improved the results for both the high and low IoU threshold in the cross-validation setting but did not yield improved results on the leaderboard. Finally, the ensemble of the two models with large patch sizes showed the best performance for an IoU threshold

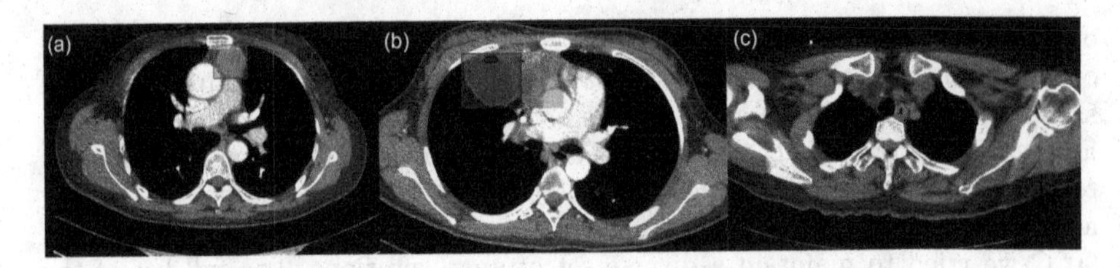

Fig. 2. Qualitative results of the final ensemble model. (a) shows an accurately detected lesion (green) (b) contains one correctly predicted lesion (green) and one false positive prediction (blue) of a lung nodule (c) shows a missed lesion (red). (Color figure online)

of 0.3 in our cross-validation and the leaderboard. The gap between the cross-validation and leaderboard results remains small for all of our models, showing good generalisation performance to unseen data. Some qualitative results for True Positive, False Positives and False Negatives from our cross-validation experiments are shown in Fig. 2.

Based on our qualitative analysis the detection of smaller lesions remains more difficult than the detection of large lesions. This might be partially caused by the downsampled data set to improve the detection performance of large lesions. Surprisingly, the model was even able to correctly predict a lung lesion which had a similar appearance as the neighboring mediastinal lesion.

4 Discussion

We presented our solution to the MELA 2022 challenge which was based on nnDetection and showed good results with little modifications to its pipeline. By using automatically generated pseudo masks, training with large patch sizes in a multi GPU setup and ensembling two models with different augmentation schemes our method achieved an excellent FROC score of 0.9922 at IoU 0.10 and 0.9880 at IoU 0.3 in our cross-validation experiments and a FROC score of 0.9897 on the leaderboard. Our submission ranked third in the MELA challenge 2022. While the detection of mediastinal lesions in this data set can already be solved with very high performance, the simultaneous detection of small and large objects remains a difficult challenge. The integration of prior knowledge into models to improve their regression performance and enhancing the post-processing procedures to improve the locatization performance of small lesions might be promising directions for future research.

Acknowledgements. Part of this work was funded by Helmholtz Imaging, a platform of the Helmholtz Incubator on Information and Data Science.

References

1. Baumgartner, M., Jäger, P.F., Isensee, F., Maier-Hein, K.H.: nnDetection: a self-configuring method for medical object detection. In: de Bruijne, M. (ed.) MICCAI 2021. LNCS, vol. 12905, pp. 530–539. Springer, Cham (2021). https://doi.org/10.1007/978-3-030-87240-3_51
2. Isensee, F., et al.: batchgenerators-a python framework for data augmentation. Zenodo. (2020)
3. Isensee, F., Jaeger, P.F., Kohl, S.A., Petersen, J., Maier-Hein, K.H.: nnU-Net: a self-configuring method for deep learning-based biomedical image segmentation. Nature Methods **18**(2), 203–211 (2021)
4. Ivantsits, M., et al.: Detection and analysis of cerebral aneurysms based on X-ray rotational angiography-the CADA 2020 challenge. Med. Image Anal. **77**, 102333 (2022)
5. Jaeger, P.F., et al.: Retina U-Net: embarrassingly simple exploitation of segmentation supervision for medical object detection. In: Machine Learning for Health Workshop, pp. 171–183. PMLR (2020)
6. Jin, L., et al.: Deep-learning-assisted detection and segmentation of rib fractures from CT scans: Development and validation of fracnet. EBioMedicine **62**, 103106 (2020)
7. Lin, T.Y., Dollár, P., Girshick, R., He, K., Hariharan, B., Belongie, S.: Feature pyramid networks for object detection. In: Proceedings of the IEEE Conference on Computer Vision and Pattern Recognition, pp. 2117–2125 (2017)
8. Setio, A.A.A., et al.: Validation, comparison, and combination of algorithms for automatic detection of pulmonary nodules in computed tomography images: the luna16 challenge. Med. Image Anal. **42**, 1–13 (2017)
9. Tang, Y.B., Yan, K., Tang, Y.X., Liu, J., Xiao, J., Summers, R.M.: Uldor: a universal lesion detector for CT scans with pseudo masks and hard negative example mining. In: 2019 IEEE 16th International Symposium on Biomedical Imaging (ISBI 2019), pp. 833–836. IEEE (2019)
10. Wu, Y., He, K.: Group normalization. In: Proceedings of the European Conference on Computer Vision (ECCV), pp. 3–19 (2018)
11. Zlocha, M., Dou, Q., Glocker, B.: Improving RetinaNet for CT lesion detection with dense masks from weak RECIST labels. In: Shen, D. (ed.) MICCAI 2019. LNCS, vol. 11769, pp. 402–410. Springer, Cham (2019). https://doi.org/10.1007/978-3-030-32226-7_45

Author Index

Y. Xiao et al. (Eds.): CuRIOUS/KiPA/MELA 2022, LNCS 13648, p. 87, 2023.
https://doi.org/10.1007/978-3-031-27324-7

Printed in the United States
by Baker & Taylor Publisher Services